M000249573

Magnetic Resonance Imaging

Principles, Methods, and Techniques

Magnetic Resonance Imaging

Principles, Methods, and Techniques

Perry Sprawls, Ph.D., FACR, FAAPM

Professor
Department of Radiology
Emory University
Atlanta, Georgia

Medical Physics Publishing
Madison, Wisconsin

Copyright 2000 by Perry Sprawls
All rights reserved.

ISBN 0-944838-97-9

05   04   03   02   01   5   4   3   2   1

Library of Congress Cataloging-in-Publication Data

Sprawls, Perry.
    Magnetic resonance imaging : principles, methods, and techniques / Perry Sprawls.
        p. ; cm.
    Includes index.
    ISBN 0-944838-97-9
    1. Magnetic resonance imaging. I. Title.
    [DNLM: 1. Magnetic Resonance Imaging--methods. 2. Health Physics. WN 185
S767m 2000]
    RC78.7.N83 S68 2000
    616.07'548--dc21
                                                                    00-034862

Perry Sprawls grants permission for photocopying for limited personal use and making slides for educational presentations if the source is fully acknowledged. This consent does not extend to other kinds of copying, such as copying for general distribution, for advertising or promotional purposes, for creating new collective works, or for resale. For information, address Perry Sprawls, 1476 Leafview Road, Decatur, GA 30033 or sprawls@emory.edu.

Every reasonable effort has been made to give factual and up-to-date information to the reader of this book. However, because of the possibility of human error and the potential for change in the medical sciences, the author, publisher, and any other persons involved in the publication of this book cannot assume responsibility for the validity of all materials or for the consequences of their use.

Medical Physics Publishing
4513 Vernon Boulevard
Madison, WI 53705-4964

Printed in the United States of America

Dedication

Even though the words and images in this book come from my hand, they are the results of years of learning and experience that have been contributed to by many others. It is with gratitude and respect that I dedicate this book to two persons who through their efforts, guidance, and support have made this possible.

Louise Bignon, *Inspiring teacher of English and Latin*

As my high school teacher, Mrs. Bignon not only guided my learning of the vocabulary, structure, and rules of the languages, but also the art of using language to effectively communicate with others. She shared with us the joy of knowing languages and using them for intellectual growth and self-development. It is the inspiration that she provides for living a meaningful life, enriched with language skills, that is appreciated. Even beyond the classroom, she demonstrates that one's individual talents take on their true value when they are used to help and bring out the best in others. Years ago, when she was teaching many (like myself) how to write, her dream was to someday write a book. I hope that this book in some small way contributes to that dream coming true.

Heinz Stephens Weens, M.D., *Pioneering and visionary radiologist*

Dr. Weens created the Emory University radiology-training program in 1946 and became Chairman of the department a year later. During his long tenure he was a leader in developing many radiological procedures, including cardiac catheterization and 35mm cineradiography, as well as many innovations in the technology of medical imaging and radiation therapy. It was my good fortune to begin my medical imaging career and to work many years under his direction. He convinced me of the need for physicists working in medical imaging and continued to guide and support my work in that direction. He had a unique vision of how new developments in science and technology could be applied to radiology. In the early 1980s he came to me and said, "Perry, there is this technique of NMR that is going to change the way we do radiology, please give the department a seminar on it." My research and study for that seminar was the seed that has grown over the years of MRI development and my work in the field that has resulted in this book.

Contents

2

Magnetic Resonance Imaging System Components

3

Nuclear Magnetic Resonance

4 Tissue Magnetization And Relaxation

8   Selective Signal Suppression

9   Spatial Characteristics of the Magnetic Resonance Image

13   Functional Imaging

14   Image Artifacts

15 MRI Safety

Preface

Magnetic resonance imaging (MRI) is a major medical diagnostic tool. It makes it possible to visualize and analyze a variety of tissue characteristics, blood flow and distribution, and several physiologic and metabolic functions. Much of this power comes from the ability to adjust the imaging process to be especially sensitive to each of the characteristics being evaluated. Think of it as a multipurpose imaging and analytical procedure that can be configured and optimized to provide answers to a wide range of clinical questions for virtually all parts and systems of the body.

Each imaging procedure is guided by a protocol consisting of a selection from a choice of imaging methods, selected values for a large number of imaging parameters or factors associated with the specific method, and the application of a variety of techniques to optimize image quality and acquire the images in the shortest time consistent with other procedure requirements.

Even though many imaging protocols are preprogrammed into modern MRI systems, maximum performance and benefit requires a highly educated and trained staff to conduct the procedures and to interpret the results.

The physicians who are requesting, supervising, and interpreting the MR examinations require knowledge of MRI principles, methods, and techniques to select appropriate imaging methods and techniques and to understand the basis for the clinical information conveyed in the images.

The technologists who perform the examinations need a good knowledge and understanding of the total process so that they can select and modify protocols as necessary, monitor and optimize image quality, and provide for patient and staff safety.

Medical physicists who provide support to the clinical activities with respect to image quality and procedure optimization, and conduct educational activities for other medical professionals must also have a broad knowledge at the practical and applied levels.

This book is designed to meet the needs of all who play a role in the MR imaging process. First, it develops the very important concepts of the physical principles on which MR imaging is based. It then builds an understanding of the various methods and techniques that are the heart of each imaging procedure. It gives special emphasis to image quality and the associated issues of optimizing protocols. Safety concerns are addressed in order to have an informed staff who can take a realistic approach to reducing risk and increasing patient comfort and acceptance.

The objective of this book is to help all of us obtain maximum performance and benefit from the advanced and sophisticated MR technology that is available today. Humans with the knowledge of how to apply the various imaging options to the wide range of clinical needs is, and will continue to be, a vital link in the total MR imaging process.

Acknowledgments

Tom Dixon, Ph.D., *for years of stimulating discussions on the physics and techniques of MRI, the review of this manuscript, and the many helpful suggestions that have been incorporated.*

Margaret Nix and Tammy Mann, *for manuscript preparation.*

Jack Peterson, Ph.D., *for technical and editorial contributions.*

Mind Maps

Mind Maps

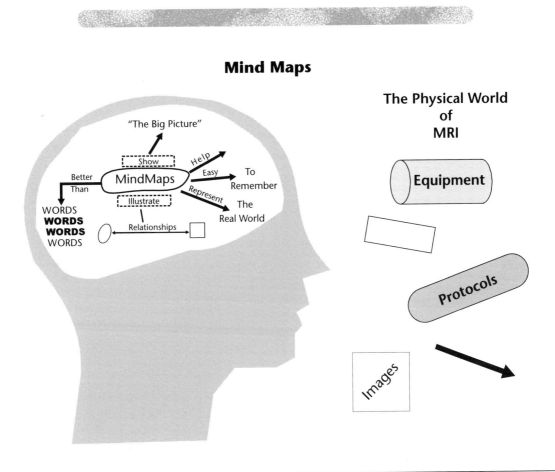

MRI is not a difficult subject to learn. It is, however, a somewhat complex topic because there are many different parts of the total MR process. A useful knowledge of MRI does not consist of memorized definitions and facts. It consists of a visual representation of the MR process in our minds. It is this type of knowledge that we can use to produce and understand images. This book is designed to help you develop the concepts of the various aspects of MRI and to see how these concepts can be applied to practical imaging situations.

Each chapter is summarized with a mind map to help organize the many individual concepts. A mind map is an effective method for showing relationships and developing a comprehensive mental picture of a topic. It is a good way to provide some organization to the complexity of a topic such as MRI. Mind maps are excellent tools for study and review. The best mind maps are often the ones that each of us develops for our own use. You are encouraged to add to and modify the printed maps and to develop mind maps of your own.

1

Magnetic Resonance Image Characteristics

Introduction And Overview

Magnetic resonance imaging (MRI) is a medical imaging process that uses a magnetic field and radio frequency (RF) signals to produce images of anatomical structures, of the presence of disease, and of various biological functions within the human body. MRI produces images that are distinctly different from the images produced by other imaging modalities. A primary difference is that the MRI process can selectively image several different tissue characteristics. A potential advantage of this is that if a pathologic process does not alter one tissue characteristic and produce contrast, it might be visible in an image because of its effect on other characteristics. This causes the MRI process to be somewhat more complex than most imaging methods. In order to optimize an MRI procedure for a specific clinical examination, the user must have a good knowledge of the characteristics of the magnetic resonance (MR) image and how those characteristics can be controlled.

In this chapter we will develop a basic knowledge and overview of the MR image, how the image relates to specific tissue characteristics, and how image quality characteristics can be controlled.

The MR Image

The MR image displays certain physical characteristics of tissue. Let us now use Figure 1-1 to identify these characteristics and to see how they are related.

The MR image is a display of RF signals that are emitted by the tissue during the image acquisition process. The source of the signals is a condition of magnetization that is produced in the tissue when the patient is placed in the strong magnetic field. The tissue magnetization depends on the presence of magnetic nuclei. The specific physical characteristic of tissue or fluid that is visible in the image depends on how the magnetic field is being changed during the acquisition process. An image acquisition consists of an acquisition cycle, like a heartbeat, that is repeated many times. During each cycle the tissue magnetization is forced through a series of changes. As we will soon learn in much more detail, all tissues and fluids do not progress through these changes at the same rate. It is the level of magnetization that is present at a special "picture snapping time" at the end of each cycle that determines the intensity of the RF signal produced and the resulting tissue brightness in the image.

MR images are generally identified with specific tissue characteristics or blood conditions

MAGNETIC RESONANCE IMAGE TYPES

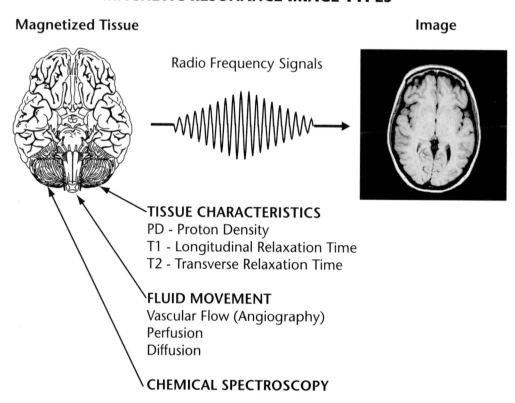

Magnetized Tissue

Radio Frequency Signals

Image

TISSUE CHARACTERISTICS
PD - Proton Density
T1 - Longitudinal Relaxation Time
T2 - Transverse Relaxation Time

FLUID MOVEMENT
Vascular Flow (Angiography)
Perfusion
Diffusion

CHEMICAL SPECTROSCOPY

Figure 1-1. Physical characteristics of tissue and fluid movement that can be displayed in the magnetic resonance image. MRI can also provide certain chemical information by applying spectroscopy analysis to the RF signals emitted by the tissue.

that are the predominant source of contrast. These characteristics determine the level of tissue magnetization and contrast present at the time the "picture is snapped." The equipment operator, who sets the imaging protocol, determines the type of image that is to be produced by adjusting various imaging factors.

The characteristics that can be used as a source of image contrast fall into three rather distinct categories. The first, and most widely used, category is the magnetic characteristics of tissues. The second category is characteristics of fluid (usually blood) movement. The third category is the spectroscopic effects related to molecular structure.

At this time we will briefly introduce each of these characteristics to set the stage for the much more detailed descriptions presented later.

Tissue Characteristics and Image Types

Proton Density (PD) Images

The most direct tissue characteristic that can be imaged is the concentration or density of protons (hydrogen). In a proton density image the tissue magnetization, RF signal intensity, and image brightness are determined by the proton (hydrogen) content of the tissue. Tissues that are rich in protons will produce strong signals and have a bright appearance.

Magnetic Relaxation Times — T1 and T2 Images

During an MRI procedure the tissue magnetization is cycled by flipping it into an unstable condition and then allowing it to recover. This recovery process is known as *relaxation*. The time required for the magnetization to relax varies from one type of tissue to another. The relaxation times can be used to distinguish (i.e., produce contrast) among normal and pathologic tissues.

Each tissue is characterized by two relaxation times: Tl and T2. Images can be created in which either one of these two characteristics is the predominant source of contrast. It is usually not possible to create images in which one of the tissue characteristics (e.g., PD, T1, or T2) is the only pure source of contrast. Typically, there is a mixing or blending of the characteristics but an image will be more heavily *weighted* by one of them. When an image is described as a T1-weighted image, this means that T1 is the predominant source of contrast but there is also some possible contamination from the PD and T2 characteristics.

Fluid Movement and Image Types

Vascular Flow

The MRI process is capable of producing images of flowing blood without the use of contrast media. Although flow effects are often visible in all types of images, it becomes the predominant source of contrast in images produced specifically for vascular or angiographic examinations as described in Chapter 12.

Perfusion and Diffusion

It is possible to produce images that show both perfusion and diffusion within tissue. These require specific imaging methods and are often characterized as functional imaging.

Spectroscopic and Chemical Shift

The frequency of the RF signals emitted by tissue is affected to a small degree by the size and characteristics of the molecules containing the magnetic nuclei. These differences in frequencies, the chemical shift, can be displayed in images. It is also the basis of MR spectroscopy. Spectroscopy is the process of using magnetic resonance to analyze the chemical composition of tissue. Spectroscopy makes use of the fact that different molecular structures have different resonant frequencies. Typically, the MR signals from a tissue specimen are sorted and displayed on a frequency scale. The signals from different chemical compounds will appear as peaks along the frequency scale. This leads to their identity and measure of relative abundance.

What Do You See In An MR Image?

We have discovered that an MR image can display a variety of tissue and body fluid characteristics. However, there are several physical characteristics that form the link between the image and the tissue characteristics described above. Understanding this link gives us a better appreciation of how the tissue characteristics are made visible. We will use Figure 1-2 to develop the link.

Radio Frequency Signal Intensity

The first thing we see in an image is RF signal intensity emitted by the tissues. Bright areas in the image correspond to tissues that emit high signal intensity. There are also areas in an image that appear as dark voids because no signals are produced. Between these two extremes there will be a range of signal intensities and shades of gray that show contrast or differences among the various tissues.

Let us now move deeper into the imaging process and discover the relationship between RF signal intensity and other characteristics.

Tissue Magnetization

The condition within the tissue that produces the RF signal is *magnetization*. At this point we will use an analogy to radioactive nuclide

imaging. In nuclear medicine procedures it is the presence of radioactivity in the tissues that produces the radiation. In MRI it is the magnetization within the tissues that produces the RF signal radiation displayed in the image. Therefore, when we look at an MR image, we are seeing a display of magnetized tissue.

We will soon discover that tissue becomes magnetized when the patient is placed in a strong magnetic field. However, all tissues are not magnetized to the same level. During the imaging process the tissue magnetization is cycled through a series of changes, but all tissues do not change at the same rate. It is this difference in rates of change of the magnetization that makes the tissues different and produces much of the useful contrast. This will be described in much more detail later when we will learn that these rates of change are described as magnetic relaxation times, T1 and T2.

It is the level of magnetization at specific "picture snapping" times during the imaging procedure that determines the intensity of the resulting RF signal and image brightness. The MR image is indeed an image of magnetized tissue. Tissues or other materials that are not adequately magnetized during the imaging procedure will not be visible in the image.

Protons (Magnetic Nuclei)

The next thing we see is an image of protons that are the nuclei of hydrogen atoms. That is why an MRI procedure is often referred to as proton imaging.

The magnetization of tissue, which produces the RF signals, comes from protons that are actually small magnets (magnetic nuclei) present in the tissue. These small magnets are actually the nuclei of certain atoms that have a special magnetic property called a *magnetic moment*. Not all chemical substances have an adequate abundance of magnetic nuclei.

WHAT DO YOU SEE IN AN MR IMAGE?

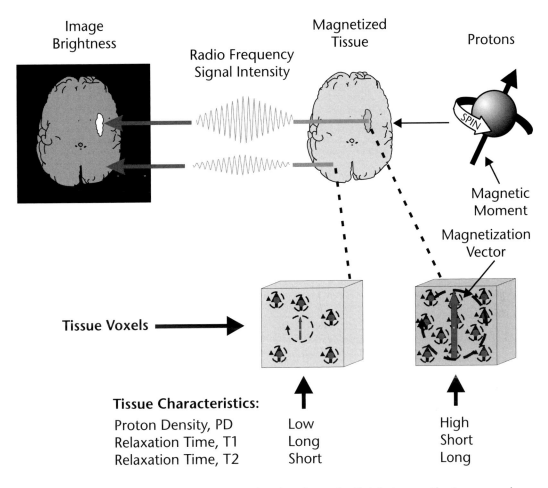

Figure 1-2. The physical characteristics that form the link between the image and the three tissue characteristics.

Hydrogen

The only substance found in tissue that has an adequate concentration of magnetic nuclei to produce good images is hydrogen. The nucleus of a hydrogen atom is a single proton. Therefore, the MR image is an image of hydrogen. When tissue that contains hydrogen (small magnetic nuclei), i.e., protons, is placed in a strong magnetic field, some of the protons line up in the same direction as the magnetic field. This alignment produces the magnetization in the tissue, which then produces the RF signal. If a tissue does not have an adequate concentration of molecules containing hydrogen, it will not be visible in an MR image.

Tissue Characteristics

As we have moved deeper into the imaging process we arrive again at the three tissue characteristics: PD, T1, and T2. It is these characteristics that we want to see because they give us valuable information about the tissues. These characteristics become visible because each one has an effect on the level of magnetization that is present at the picture snapping time in each imaging cycle. At this time we will briefly describe the effect of each and then develop the process in more detail in Chapters 4 and 5.

PD (Proton Density)

PD has a very direct effect on tissue magnetization and the resulting RF signal and image brightness. That is because the magnetization is produced by the protons. Therefore, a tissue with a high PD can reach a high level of magnetization and produce an intense signal.

T1

When the imaging protocol is set to produce a T1-weighted image, it is the tissues with the short T1 values that produce the highest magnetization and are the brightness in the image.

T2

When the imaging protocol is set to produce a T2-weighted image, it is the tissues with the long T2 values that are the brightest. This is because they have a higher level of magnetization at the picture snapping time.

Spatial Characteristics

Figure 1-3 illustrates the basic spatial characteristics of the MR image. MRI is basically a tomographic imaging process, although there are some procedures, such as angiography, in which a complete anatomical volume will be displayed in a single image. The protocol for the acquisition process must be set up to produce the appropriate spatial characteristics for a specific clinical procedure. This includes such factors as the number of slices, slice orientation, and the structure within each individual slice.

Slices

A typical examination will consist of at least one set of contiguous slices. In most cases the entire set of slices is acquired simultaneously. However, the number of slices in a set can be limited by certain imaging factors and the amount of time allocated to the acquisition process.

The slices can be oriented in virtually any plane through the patient's body. The major restriction is that images in the different planes cannot generally be acquired simultaneously. For example, if both axial and sagittal images are required, the acquisition process must be repeated. However, there is the possibility of acquiring 3-D data from a large volume of tissue and then reconstructing slices in the different planes, as will be described in Chapter 9.

Voxels

Each slice of tissue is subdivided into rows and columns of individual volume elements, or voxels. The size of a voxel has a significant effect on image quality. It is controlled by a combination of protocol factors as described in Chapter 10 and should be adjusted to an optimum size for each type of clinical examination. Each voxel is an independent source of RF signals. That is why voxel size is a major consideration in each image acquisition.

Image Pixels

The image is also divided into rows and columns of picture elements, or pixels. In general, an image pixel represents a corresponding voxel of tissue within the slice. The brightness of an image pixel is determined by the intensity of the RF signal emitted by the tissue voxel.

SPATIAL CHARACTERISTICS

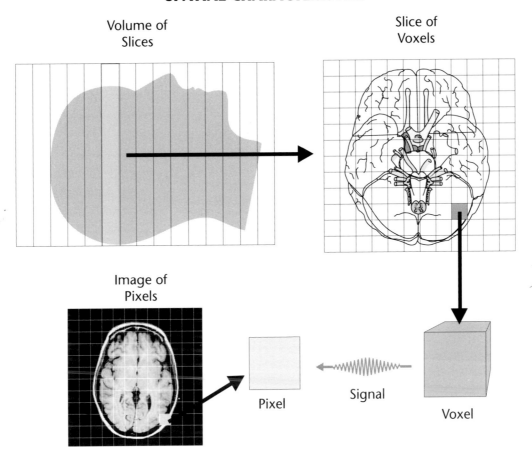

Figure 1-3. The spatial characteristics of MR images.

Control Of Image Characteristics

The operator of an MRI system has tremendous control over the characteristics and the quality of the images that are produced. The five basic image quality characteristics are represented in Figure 1-4. Each of these image characteristics is affected by a combination of the imaging factors that make up the acquisition protocol.

Not all types of clinical procedures require images with the same characteristics. Therefore, the primary objective is to use an imaging protocol in which the acquisition process is optimized for a specific clinical requirement.

Although each of the image characteristics will be considered in detail in later chapters, we will introduce them here.

Contrast Sensitivity

Contrast sensitivity is the ability of an imaging process to produce an image of objects or tissues in the body that have relatively small physical differences or inherent contrast. The contrast that is to be imaged is in the form of

MAGNETIC RESONANCE IMAGE QUALITY CHARACTERISTICS

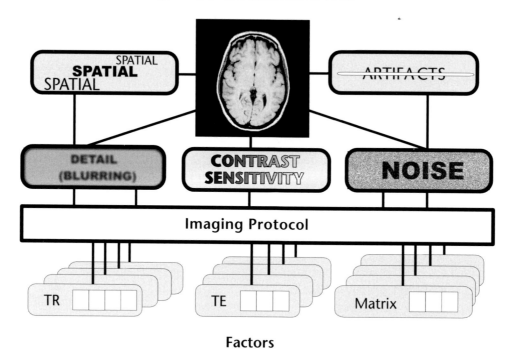

Figure 1-4. Image quality characteristics that can be controlled by the selection of protocol factors.

some specific physical characteristic. In x-ray imaging, including CT (computed tomography), difference in physical density is a principle source of contrast. One of the major advantages of MRI is that it has a high contrast sensitivity for visualizing differences among the tissues in the body because there are several sources of contrast; that is, it has the ability to image a variety of characteristics (PD, T1, T2) as described previously. Also, there is usually much greater variation among these characteristics than among the tissue density values that are the source of contrast for x-ray imaging. If a certain pathologic condition does not produce a visible change in one characteristic, there is the possibility that it will be visible by imaging some of the other characteristics.

Even though MRI has high contrast sensitivity relative to most of the other imaging modalities, it must be optimized for each clinical procedure. This includes the selection of the characteristics, or sources of contrast, that are to be imaged and then adjusting the protocol factors so that the sensitivity to that specific characteristic is optimized. This is illustrated in Figure 1-5.

Detail

A distinguishing characteristic of every imaging modality is its ability to image small objects and structures within the body. Visibility of anatomical detail (sometimes referred to as spatial resolution) is limited by the blurring that occurs during the imaging process. All medical imaging methods produce images with some blurring but not to the same extent. The blurring in MRI is greater than in radiography. Therefore, MRI cannot image small structures that are visible in conventional radiographs.

In MRI, like all modalities, the amount of blurring and the resulting visibility of detail can be adjusted during the imaging process. Figure 1-6 shows images with different levels of blurring and visibility of detail. The protocol factors that are used to adjust detail and the associated issues in their optimization will be discussed in Chapter 10.

Noise

Visual noise is a major issue in MRI. The presence of noise in an image reduces its quality, especially by limiting the visibility of low contrast objects and differences among tissues. Figure 1-7 shows images with different levels of

visual noise. Most of the noise in MR images is the result of a form of random, unwanted RF energy picked up from the patient's body.

The amount of noise can generally be controlled through a combination of factors as described in Chapter 10. However, many of these factors involve compromises with other characteristics.

Artifacts

Artifacts are undesirable objects, such as streaks and spots, that appear in images which do not directly represent an anatomical structure. They are usually produced by certain interactions of the patient's body or body functions (such as motion) with the imaging process.

There is a selection of techniques that can be used to reduce the presence of artifacts. These will be described in Chapter 14.

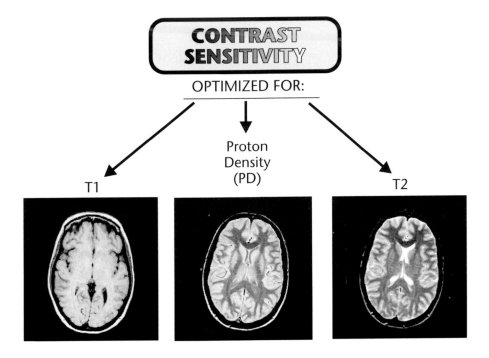

Figure 1-5. The images produced when the contrast sensitivity is optimized for each of the three specific tissue characteristics.

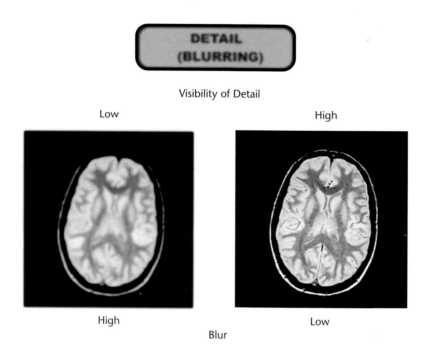

DETAIL
(BLURRING)

Visibility of Detail

Low   High

High   Low

Blur

Figure 1-6. Images with different levels of blurring and visibility of anatomical detail.

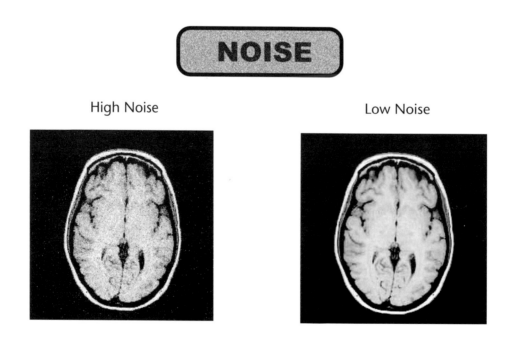

NOISE

High Noise   Low Noise

Figure 1-7. Images with different levels of visual noise.

Spatial

The general spatial characteristics of the MR image were described previously. However, when setting up an imaging protocol the spatial characteristics must be considered in the general context of image quality. As we will discover later, voxel size plays a major role in determining both image detail and image noise.

Image Acquisition Time

When considering and adjusting MR image quality, attention must also be given to the time required for the acquisition process. In general, several aspects of image quality, such as detail and noise, can be improved by using longer acquisition times.

Protocol Optimization

An optimum imaging protocol is one in which there is a proper balance among the image quality characteristics described above and also a balance between overall image quality and acquisition time.

The imaging protocol that is used for a specific clinical examination has a major impact on the quality of the image and the visibility of anatomical structures and pathologic conditions.

Therefore, the users of MRI must have a good knowledge of the imaging process and the protocol factors and know how to set them to optimize the image characteristics.

The overall process of optimizing protocols will be described in Chapter 11.

Mind Map Summary
Magnetic Resonance Image Characteristics

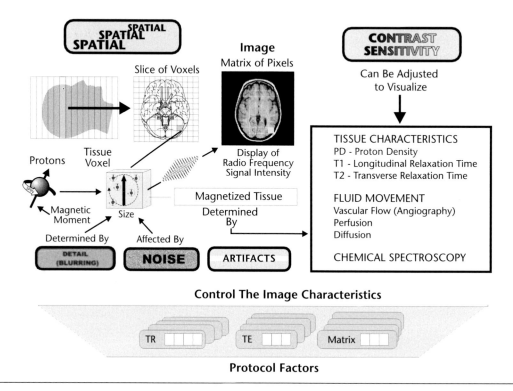

Control The Image Characteristics

Protocol Factors

The magnetic resonance image is a display of radio frequency signal intensities that are emitted by magnetized tissue during the imaging process. The tissue becomes magnetized because it contains protons that are the magnetic nuclei of hydrogen atoms. When placed in the strong magnetic field, some of the protons align with the field producing the tissue magnetization. The level of magnetization at the time during the procedure when the "picture is snapped" is determined by a variety of tissue and fluid movement characteristics. By adjusting the imaging process it is possible to produce images in which these various characteristics are the principal sources of contrast.

An advantage of MRI is the ability to selectively image a variety of tissue and fluid characteristics. If a specific pathologic condition is not visible when viewing one characteristic, there is the possibility of seeing it by imaging some of the other characteristics.

During the imaging procedure a section of the patient's body is divided first into slices, and the slices are divided into a matrix of voxels. Each voxel is an independent RF signal source. Voxel size can be adjusted and is what determines image detail and also affects image noise.

The five major image quality characteristics—contrast sensitivity, detail, noise, artifacts, and spatial—can be controlled to a great extent by the settings of the various protocol factors.

MRI is a powerful diagnostic tool because the process can be optimized to display a wide range of clinical conditions. However, maximum benefit requires a staff with the knowledge to control the process and interpret the variety of images.

2

Magnetic Resonance Imaging System Components

Introduction And Overview

The MRI system consists of several major components, as shown in Figure 2-1. At this time we will introduce the components and indicate how they work together to create the MR image. The more specific details of the image forming process will be explained in later chapters.

The heart of the MRI system is a large magnet that produces a very strong magnetic field. The patient's body is placed in the magnetic field during the imaging procedure. The magnetic field produces two distinct effects that work together to create the image.

Tissue Magnetization

When the patient is placed in the magnetic field, the tissue becomes temporarily magnetized because of the alignment of the protons, as described previously. This is a very low-level effect that disappears when the patient is removed from the magnetic field. The ability of MRI to distinguish between different types of tissue is based on the fact that different tissues, both normal and pathologic, will become magnetized to different levels or will change their levels of magnetization (i.e., relax) at different rates.

THE MRI SYSTEM

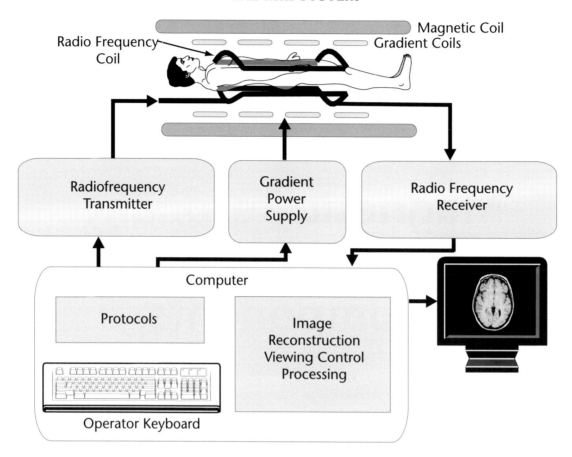

Figure 2-1. The major components of the Magnetic Resonance Imaging System.

Tissue Resonance

The magnetic field also causes the tissue to "tune in" or resonate at a very specific radio frequency. That is why the procedure is known as *magnetic resonance imaging*. It is actually certain nuclei, typically protons, within the tissue that resonate. Therefore, the more comprehensive name for the phenomenon that is the basis of both imaging and spectroscopy is *nuclear magnetic resonance* (NMR).

In the presence of the strong magnetic field the tissue resonates in the RF range. This causes the tissue to function as a tuned radio receiver and transmitter during the imaging process.

The production of an MR image involves two-way radio communication between the tissue in the patient's body and the equipment.

The Magnetic Field

Figure 2-2 shows the general characteristics of a typical magnetic field. At any point within a magnetic field, the two primary characteristics are *field direction* and *field strength*.

Field Direction

It will be easier to visualize a magnetic field if it is represented by a series of parallel lines, as shown in Figure 2-2. The arrow on each line

THE MAGNETIC FIELD

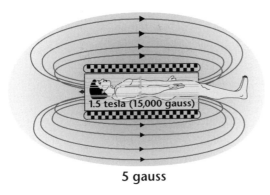

1.5 tesla (15,000 gauss)

5 gauss

Figure 2-2. The magnetic field produced by superconducting magnets.

indicates the direction of the field. On the surface of the earth, the direction of the earth's magnetic field is specified with reference to the north and south poles. The north-south designation is generally not applied to magnetic fields used for imaging. Most of the electromagnets used for imaging produce a magnetic field that runs through the bore of the magnet and parallel to the major patient axis. As the magnetic field leaves the bore, it spreads out and encircles the magnet, creating an external fringe field. The external field can be a source of interference with other devices and is usually contained by some form of shielding.

Field Strength

Each point within a magnetic field has a particular intensity, or strength. Field strength is expressed either in the units of tesla (T) or gauss (G). The relationship between the two units is that 1.0 T is equal to 10,000 G or 10 kG. At the earth's surface, the magnetic field is relatively weak and has a strength of less than 1 G. Magnetic field strengths in the range of 0.15 T to 1.5 T are used for imaging. The significance of

field strength is considered as we explore the characteristics of MR images and image quality in later chapters.

Homogeneity

MRI requires a magnetic field that is very uniform, or homogeneous with respect to strength. Field homogeneity is affected by magnet design, adjustments, and environmental conditions. Imaging generally requires a homogeneity (field uniformity) on the order of a few parts per million (ppm) within the imaging area.

High homogeneity is obtained by the process of shimming, as described later.

Magnets

There are several different types of magnets that can be used to produce the magnetic field. Each has its advantages and disadvantages.

Superconducting

Most MRI systems use superconducting magnets. The primary advantage is that a superconducting magnet is capable of producing a much stronger and stable magnetic field than the other two types (resistive and permanent) considered below. A superconducting magnetic is an electromagnet that operates in a superconducting state. A superconductor is an electrical conductor (wire) that has no resistance to the flow of an electrical current. This means that very small superconducting wires can carry very large currents without overheating, which is typical of more conventional conductors like copper. It is the combined ability to construct a magnet with many loops or turns of small wire and then use large currents that makes the strong magnetic fields possible.

There are two requirements for superconductivity. The conductor or wire must be fabricated from a special alloy and then cooled to a very low temperature. The typical magnet consists of small niobium-titanium (Nb-Ti) wires imbedded in copper. The copper has electrical resistance and actually functions as an insulator around the Nb-Ti superconductors.

During normal operation the electrical current flows through the superconductor without dissipating any energy or producing heat. If the temperature of the conductor should ever rise above the critical superconducting temperature, the current begins to produce heat and the current is rapidly reduced. This results in the collapse of the magnetic field. This is an undesirable event known as a *quench*. More details are given in Chapter 15 on safety. Superconducting magnets are cooled with liquid helium. A disadvantage of this magnet technology is that the coolant must be replenished periodically.

A characteristic of most superconducting magnets is that they are in the form of cylindrical or solenoid coils with the strong field in the internal bore. A potential problem is that the relatively small diameter and the long bore produce claustrophobia in some patients. Superconducting magnetic design is evolving to more open patient environments to reduce this concern.

Resistive

A resistive type magnet is made from a conventional electrical conductor such as copper. The name "resistive" refers to the inherent electrical resistance that is present in all materials except for superconductors. When a current is passed through a resistive conductor to produce a magnetic field, heat is also produced. This limits this type of magnet to relatively low field strengths.

Permanent

It is possible to do MRI with a non-electrical permanent magnet. An obvious advantage is that a permanent magnet does not require either electrical power or coolants for operation. However, this type of magnet is also limited to relatively low field strengths.

Both resistive and permanent magnets are usually designed to produce vertical magnetic fields that run between the two magnetic poles, as shown in Figure 2-3. Possible advantages include a more open patient environment and less external field than superconducting magnets.

Gradients

When the MRI system is in a resting state and not actually producing an image, the magnetic field is quite uniform or homogeneous over the region of the patient's body. However, during the imaging process the field must be distorted with gradients. A gradient is just a

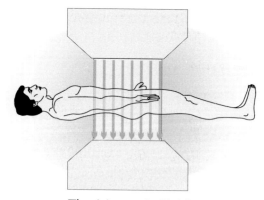

PERMANENT OR RESISTIVE MAGNETS

The Magnetic Field

Arrows Indicate Field Direction

Figure 2-3. The magnetic field produced by typical resistive or permanent magnets.

change in field strength from one point to another in the patient's body. The gradients are produced by a set of gradient coils, which are contained within the magnet assembly. During an imaging procedure the gradients are turned on and off many times. This action produces the sound or noise that comes from the magnet.

The effect of a gradient is illustrated in Figure 2-4. When a magnet is in a "resting state," it produces a magnetic field that is uniform or homogenous over most of the patient's body. In this condition there are no gradients in the field. However, when a gradient coil is turned on by applying an electric current, a gradient or variation in field strength is produced in the magnetic field.

Gradient Orientation

The typical imaging magnet contains three separate sets of gradient coils. These are oriented so that gradients can be produced in the three orthogonal directions (often designated as the x, y, and z directions). Also, two or more of the gradient coils can be used together to produce a gradient in any desired direction.

Gradient Functions

The gradients are used to perform many different functions during the image acquisition process. It is the gradients that create the spatial characteristics by producing the slices and voxels that will be described in Chapter 9. The entire family of gradient echo imaging

A MAGNETIC FIELD GRADIENT

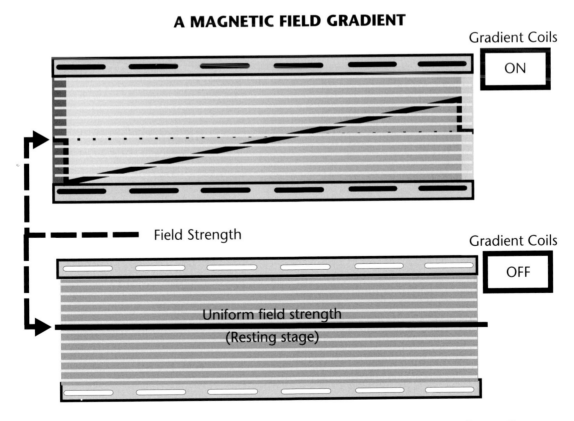

Figure 2-4. A magnetic field gradient produced by a current in the gradient coil.

methods uses a gradient to produce the echo event and signal which will be described in Chapter 7. Gradients are also used to produce one type of image contrast (phase contrast angiography) for vascular imaging, as will be described in Chapter 12, and in the functional imaging methods described in Chapter 13. Gradients also are used as part of some of the techniques to reduce image artifacts, as will be described in Chapter 14.

Gradient Strength

The strength of a gradient is expressed in terms of the change in field strength per unit of distance. The typical units are millitesla per meter (mT/m). The maximum gradient strength that can be produced is a design characteristic of a specific imaging system. High gradient strengths of 20 mT/m or more are required for the optimum performance of some imaging methods.

Risetime and Slew-Rate

For certain functions it is necessary for the gradient to be capable of changing rapidly. The *risetime* is the time required for a gradient to reach its maximum strength. The *slew-rate* is the rate at which the gradient changes with time. For example, a specific gradient system might have a risetime of 0.20 milliseconds (msec) and a slew-rate of 100 mT/m/msec.

Eddy Currents

Eddy currents are electrical currents that are induced or generated in metal structures or conducting materials that are within a changing magnetic field. Since gradients are strong, rapidly changing magnetic fields, they are capable of producing undesirable eddy currents in some of the metal components of the magnet assembly. This is undesirable because

the eddy currents create their own magnetic fields that interfere with the imaging process.

Gradients are designed to minimize eddy currents either with special gradient shielding or electrical circuits that control the gradient currents in a way that compensates for the eddy-current effects.

Shimming

One of the requirements for good imaging is a homogeneous magnet field. This is a field in which there is a uniform field strength over the image area. Shimming is the process of adjusting the magnetic field to make it more uniform.

Inhomogeneities are usually produced by magnetically susceptible materials located in the magnetic field. The presence of these materials produces distortions in the magnetic field that are in the form of inhomogeneities. This can occur in both the internal and external areas of the field. Each time a different patient is placed in the magnetic field, some inhomogeneities are produced. There are many things in the external field, such as building structures and equipment, that can produce inhomogeneities. The problem is that when the external field is distorted, these distortions are also transferred to the internal field where they interfere with the imaging process. Inhomogeneities produce a variety of problems that will be discussed later.

It is not possible to eliminate all of the sources of inhomogeneities. Therefore, shimming must be used to reduce the inhomogeneities. This is done in several ways. When a magnet is manufactured and installed, some shimming might be done by placing metal shims in appropriate locations. Magnets also contain a set of shim coils. Shimming is produced by adjusting the electrical currents in these coils. General shimming is done by the engineers when a magnet is installed or serviced. Additional shimming is done for individual patients. This is often done automatically by the system.

Magnetic Field Shielding

The external magnetic field surrounding the magnet is the possible source of two types of problems. One problem is that the field is subject to distortions by metal objects (building structures, vehicles, etc.) as described previously. These distortions produce inhomogeneities in the internal field. The second problem is that the field can interfere with many types of electronic equipment such as imaging equipment and computers.

It is a common practice to reduce the size of the external field by installing shielding as shown in Figure 2-5. The principle of magnetic field shielding is to provide a more attractive return path for the external field as it passes from one end of the magnetic field to the other. This is possible because air is not a good magnetic field conductor and can be replaced by more conductive materials, such as iron. There are two types of shielding: *passive* and *active*.

Passive Shielding

Passive shielding is produced by surrounding the magnet with a structure consisting of relatively large pieces of ferromagnetic materials such as iron. The principle is that the

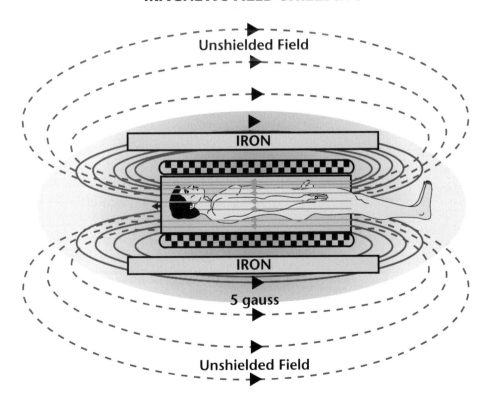

MAGNETIC FIELD SHIELDING

Figure 2-5. The principle of magnetic field shielding.

ferromagnetic materials are a more attractive path for the magnetic field than the air. Rather than expanding out from the magnet, the magnetic field is concentrated through the shielding material located near the magnet as shown in Figure 2-5. This reduces the size of the field.

Active Shielding

Active shielding is produced by additional coils built into the magnet assembly. They are designed and oriented so that the electrical currents in the coils produce magnetic fields that oppose and reduce the external magnetic field.

The Radio Frequency System

The radio frequency (RF) system provides the communications link with the patient's body for the purpose of producing an image. All medical imaging modalities use some form of radiation (e.g., x-ray, gamma-ray, etc.) or energy (e.g., ultrasound) to transfer the image from the patient's body.

The MRI process uses RF signals to transmit the image from the patient's body. The RF energy used is a form of non-ionizing radiation. The RF pulses that are applied to the patient's body are absorbed by the tissue and converted to heat. A small amount of the energy is emitted by the body as signals used to produce an image. Actually, the image itself is not formed within and transmitted from the body. The RF signals provide information (data) from which the image is reconstructed by the computer. However, the resulting image is a display of RF signal intensities produced by the different tissues.

RF Coils

The RF coils are located within the magnet assembly and relatively close to the patient's body. These coils function as the antennae for both transmitting signals to and receiving

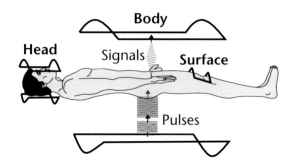

THE RF COILS

Figure 2-6. The three types of RF coils (body, head, and surface) that are the antennae for transmitting pulses and receiving signals from the patient's body.

signals from the tissue. There are different coil designs for different anatomical regions (shown in Figure 2-6). The three basic types are body, head, and surface coils. The factors leading to the selection of a specific coil will be considered in Chapter 10. In some applications the same coil is used for both transmitting and receiving; at other times, separate transmitting and receiving coils are used.

Surface coils are used to receive signals from a relatively small anatomical region to produce better image quality than is possible with the body and head coils. Surface coils can be in the form of single coils or an array of several coils, each with its own receiver circuit operated in a *phased array* configuration. This configuration produces the high image quality obtained from small coils but with the added advantage of covering a larger anatomical region and faster imaging.

Transmitter

The RF transmitter generates the RF energy, which is applied to the coils and then transmitted to the patient's body. The energy is generated as a series of discrete RF pulses. As we will see in Chapters 6, 7, and 8, the characteristics of an image are determined by the specific sequence of RF pulses.

The transmitter actually consists of several components, such as RF modulators and power amplifiers, but for our purposes here we will consider it as a unit that produces pulses of RF energy. The transmitters must be capable of producing relatively high power outputs on the order of several thousand watts. The actual RF power required is determined by the strength of the magnetic field. It is actually proportional to the square of the field strength. Therefore, a 1.5 T system might require about nine times more RF power applied to the patient than a 0.5 T system. One important component of the transmitter is a power monitoring circuit. That is a safety feature to prevent excessive power being applied to the patient's body, as described in Chapter 15.

Receiver

A short time after a sequence of RF pulses is transmitted to the patient's body, the resonating tissue will respond by returning an RF signal. These signals are picked up by the coils and processed by the receiver. The signals are converted into a digital form and transferred to the computer where they are temporarily stored.

RF Polarization

The RF system can operate either in a linear or a circularly polarized mode. In the circularly polarized mode, quadrature coils are used. Quadrature coils consist of two coils with a 90° separation. This produces both improved excitation efficiency by producing the same effect with half of the RF energy (heating) to the patient, and a better signal-to-noise ratio for the received signals.

RF Shielding

RF energy that might be in the environment could be picked up by the receiver and interfere with the production of high quality images. There are many sources of stray RF energy, such as fluorescent lights, electric motors, medical equipment, and radio communications devices. The area, or room, in which the patient's body is located must be shielded against this interference.

An area can be shielded against external RF signals by surrounding it with an electrically conducted enclosure. Sheet metal and copper screen wire are quite effective for this purpose.

The principle of RF shielding is that RF signals cannot enter an electrically conductive enclosure. The thickness of the shielding is not a factor—even thin foil is a good shield. The important thing is that the room must be completely enclosed by the shielding material without any holes. The doors into imaging rooms are part of the shielding and should be closed during image acquisition.

Computer Functions

A digital computer is an integral part of an MRI system. The production and display of an MR image is a sequence of several specific steps that are controlled and performed by the computer.

Acquisition Control

The first step is the acquisition of the RF signals from the patient's body. This acquisition process consists of many repetitions of an imaging cycle. During each cycle a sequence of RF pulses is transmitted to the body, the gradients are activated, and RF signals are collected. Unfortunately, one imaging cycle does not produce enough signal data to create an image. Therefore, the imaging cycle must be repeated many times to form an image. The time required to acquire images is determined by the duration of the imaging cycle or cycle repetition time—an adjustable factor known as TR—and the number of cycles. The number of cycles used is related to image quality. More cycles generally produce better images. This will be described in much more detail in Chapters 10 and 11.

Protocols stored in the computer control the acquisition process. The operator can select from many preset protocols for specific clinical procedures or change protocol factors for special applications.

Image Reconstruction

The RF signal data collected during the acquisition phase is not in the form of an image. However, the computer can use the collected data to create or "reconstruct" an image. This is a mathematical process known as a Fourier transformation that is relatively fast and usually does not have a significant effect on total imaging time.

Image Storage and Retrieval

The reconstructed images are stored in the computer where they are available for additional processing and viewing. The number of images that can be stored—and available for immediate display—depends on the capacity of the storage media.

Viewing Control and Post Processing

The computer is the system component that controls the display of the images. It makes it possible for the user to select specific images and control viewing factors such as windowing (contrast) and zooming (magnification).

In many applications it is desirable to process the reconstructed images to change their characteristics, to reformat an image or set of images, or to change the display of images to produce specific views of anatomical regions.

These post-processing (after reconstruction) functions are performed by a computer. In some MRI systems some of the post processing is performed on a work-station computer that is in addition to the computer contained in the MRI system.

Mind Map Summary
Magnetic Resonance Imaging System Components

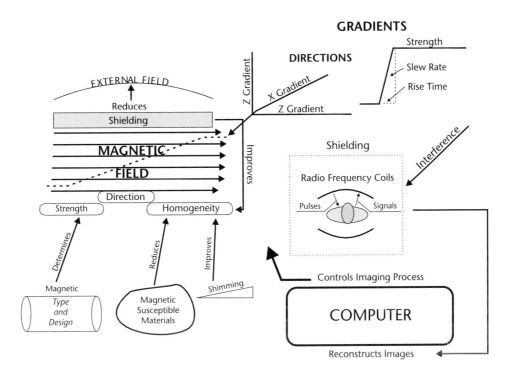

The magnetic resonance imaging system consists of several major components that function together to produce images. During the image acquisition process the patient's body is placed in a strong magnetic field. At each point, the magnetic field has a specific direction. This direction is used as a reference for expressing the direction of tissue magnetization. The strength of a magnetic field is determined by the type and design of the magnet. Superconducting magnets can produce strong magnetic fields. Resistive and permanent magnets are limited to relatively weak field strengths. The homogeneity, or uniformity of field strength is necessary for good imaging. Homogeneity is reduced by magnetically susceptible materials that come into the field and produce distortions. This can occur in both the external field and within a patient's body. Shimming is the process of adjusting the magnetic field to make it more homogeneous. This can be achieved by passive shims that are added when a magnet is installed and with active shimming produced by adjusting the currents in the shimming coils.

Shielding of the magnetic field reduces the size and strength of the external magnetic field and also improves homogeneity by protecting from interference caused by objects in the external field area.

A gradient is an intentional variation in magnetic field strength that is produced by the gradient coils. There are three basic gradient coils that are oriented to produce gradients in the three orthogonal directions. Gradients perform several functions during the image acquisition process. An important characteristic of a gradient, especially for some advanced image procedures, is its strength and how fast it can be turned on and off.

The MRI process consists of an exchange of RF pulses and signals between the equipment and the patient's body. This is done through the RF coils that serve as the antenna for transmitting the pulses and receiving the signals. It is necessary to shield the imaging area by enclosing it in a conductive metal (copper) room to block external RF interference.

The imaging process is controlled by information stored in a computer. The protocols programmed into the computer and selected by the operator guide the imaging process and determine the characteristics of the images. The RF signals collected from the patient's body during the acquisition process are used by the computer to reconstruct the image.

3

Nuclear Magnetic Resonance

Introduction And Overview

When certain materials, such as tissue, are placed in a strong magnetic field, two things happen. The materials take on a *resonant characteristic* and they become *magnetized*. In this chapter we will consider the resonant characteristic. In Chapter 4 we will study the magnetization effect. *Resonance* means the materials can absorb and then re-radiate RF radiation at a specific frequency, like a radio receiver-transmitter, as illustrated in Figure 3-1. It is actually the nuclei of the atoms that resonate. The phenomenon is generally known as nuclear magnetic resonance (NMR). The resonant frequency of material such as tissue is typically in the RF range so that the emitted radiation is in

the form of radio signals. The specific resonant frequency is determined by three factors as shown in the illustration and will be described in detail later. The characteristics of the RF signals emitted by the material are determined by certain physical and chemical characteristics of the material. The RF signals produced by the NMR process can be displayed either in the form of images (MRI) or as a graph depicting chemical composition (MR spectroscopy).

Magnetic Nuclei

Materials that participate in the MR process must contain nuclei with specific magnetic properties. In order to interact with a magnetic field, the nuclei themselves must be

MAGNETIC RESONANCE

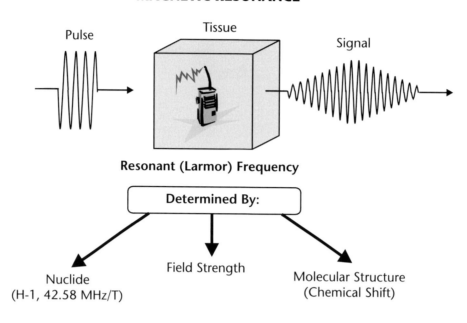

Figure 3-1. The concept of Nuclear Magnetic Resonance (NMR).

small magnets and have a magnetic property or *magnetic moment,* as shown in Figure 3-2. The magnetic characteristic of an individual nucleus is determined by its neutron-proton composition. Only certain nuclides with an odd number of neutrons and protons are magnetic. Even though most chemical elements have one or more isotopes with magnetic nuclei, the number of magnetic isotopes that might be useful for either imaging or *in vivo* spectroscopic analysis is somewhat limited. Among the nuclides that are magnetic and can participate in an NMR process, the amount of signal produced by each nuclide varies considerably.

Spins

Protons and neutrons that make up a nucleus have an intrinsic angular momentum or *spin.* Pairs of protons and neutrons align in such a way that their spins cancel. However, when there is an odd number of protons or neutrons

(odd mass numbers), some of the spins will not be canceled and the total nucleus will have a net spin characteristic. It is this spinning characteristic of a particle with an electric charge (the nucleus) that produces a magnetic property known as the *magnetic moment.*

It is for this reason that magnetic nuclei, such as protons, are often referred to as *spins.*

The magnetic property, or magnetic moment, of a nucleus has a specific direction. In Figure 3-2, the direction of the magnetic moment is indicated by an arrow drawn through the nucleus.

RF Signal Intensity

The intensity of the RF signal emitted by tissue is probably the most significant factor in determining image quality and the time required to acquire an image. This important issue is considered in Chapters 10 and 11. We now begin to introduce the factors that contribute to signal intensity.

MAGNETIC RESONANCE IMAGING

MAGNETIC PROPERTIES OF NUCLEI

Magnetic Moment   **Non-Magnetic**

Odd Mass Numbers   **Even Mass Numbers**
Hydrogen-1   Carbon-12
                                        Oxygen-16

Figure 3-2. Magnetic and non-magnetic nuclei.

During the imaging process, the body section is divided into an array of individual volume elements, or voxels. It is the signal intensity from each voxel that determines image quality. The signal is produced by the magnetic nuclei within each voxel. Therefore, signal intensity is, in general, proportional to the quantity of magnetic nuclei within an individual voxel. We now consider the factors that affect the number of magnetic nuclei within an individual voxel.

Relative Signal Strength

The relative signal strength from the various chemical elements in tissue is determined by three factors: (1) tissue concentration of the element; (2) isotopic abundance; and (3) sensitivity of the specific nuclide.

In comparison to all other nuclides, hydrogen produces an extremely strong signal. This results from its high values for each of the three contributing factors.

Of the three factors, only the concentration, or density, of the nuclei varies from point to point within an imaged section of tissue. The quantity is often referred to as *proton density* and is the most fundamental tissue

characteristic that determines the intensity of the RF signal from an individual voxel, and the resulting pixel brightness. In most imaging situations, pixel brightness is proportional to the density (concentration) of nuclei (protons) in the corresponding voxel, although additional factors, such as relaxation times, modify this relationship.

Protons in solids, such as the tabletop and bone, do not produce signals. Signals come only from protons in molecules that are free to move, as in a liquid state.

Tissue Concentration of Elements

The concentration of chemical elements in tissue covers a considerable range, depending on tissue type and such factors as metabolic or pathologic state. The concentrations of elements in tissue are in two groups. Four elements—hydrogen, carbon, nitrogen, and oxygen—typically make up at least 99% of tissue mass.

The most abundant isotopes of the four elements are hydrogen-1, carbon-12, nitrogen-14, and oxygen-16. Note that the mass number of hydrogen (1) is odd while the mass numbers of the other three (12, 14, 16) are even. Therefore, hydrogen is the only one of these four isotopes that has a strong magnetic nucleus. The nucleus of the hydrogen-1 atom is a single proton. Among all the chemical elements, hydrogen is unique in that it occurs in relatively high concentrations in most tissues, and the most abundant isotope (H-1) has a magnetic nucleus.

Other elements, such as sodium, phosphorus, potassium, and magnesium, are present in very low concentrations. Calcium is concentrated in bone or localized deposits.

Within this group of elements with low tissue concentrations are several with magnetic nuclei. These include fluorine-19, sodium-23, phosphorus-31, and potassium-39.

Isotopic Abundance

Most chemical elements have several isotopes. When a chemical element is found in a naturally occurring substance, such as tissue, most of the element is typically in the form of one isotope, with very low concentrations of the other isotopic forms. For the three elements—carbon, nitrogen, and oxygen—that have a high concentration in tissue, the magnetic isotopes are the ones with a low abundance in the natural state. These include carbon-l3, nitrogen-15, and oxygen-17.

Relative Sensitivity and Signal Strength

The signal strength produced by an equal quantity of the various nuclei also varies over a considerable range. This inherent NMR sensitivity is typically expressed relative to hydrogen-1, which produces the strongest signal of all of the nuclides. The relative sensitivities of some magnetic nuclides are shown in Table 3-1.

Table 3-1. Relative Sensitivities of Some Magnetic Nuclides

Nuclide	Sensitivity
Hydrogen-1	1.0
Fluorine-19	0.83
Sodium-23	0.093
Phosphorus-1	0.066

In summary, hydrogen has a lot going for it: 1) a high tissue concentration; 2) the most abundant isotope (H-1) is magnetic; and 3) it produces a relatively strong signal compared to an equal concentration of other nuclei. That is why hydrogen is the only element that is imaged with conventional MRI systems.

Radio Frequency Energy

During an imaging procedure, RF energy is exchanged between the imaging system and the patient's body. This exchange takes place through a set of coils located relatively close to the patient's body as we saw in Chapter 2. The RF coils are the antennae that transmit energy to and receive signals from the tissue.

Pulses

RF energy is applied to the body in several short pulses during each imaging cycle. The strength of the pulses is described in terms of the angle through which they rotate or flip the magnetic nuclei and the resulting tissue magnetization, as described later. Many imaging methods use both 90° and 180° pulses in each cycle.

Signals

At a specific time in each imaging cycle, the tissue is stimulated to emit an RF signal, which is picked up by the coils, analyzed, and used to form the image. The spin echo or gradient echo methods are generally used to stimulate signal emission. Therefore, the signals from the patient's body are commonly referred to as *echoes*.

Nuclear Magnetic Interactions

The NMR process is a series of interactions involving the magnetic nuclei, a magnetic field, and RF energy pulses and signals.

Nuclear Alignment

Recall that a magnetic nucleus is characterized by a magnetic moment. The direction of the magnetic moment is represented by a small arrow passing through the nucleus. If we think of the nucleus as a small conventional magnet, the magnetic moment arrow corresponds to the south pole-north pole direction of the magnet.

In the absence of a strong magnetic field, magnetic moments of nuclei are randomly oriented in space. Many nuclei in tissue are not in a rigid structure and are free to change direction. In fact, nuclei are constantly tumbling, or changing direction, because of thermal activity within the material; in this case, tissue.

When a material containing magnetic nuclei is placed in a magnetic field, the nuclei experience a torque that encourages them to align with the direction of the field. In the human body, however, thermal energy agitates the nuclei and keeps most of them from aligning parallel to the magnetic field. The number of nuclei that do align with the magnetic field is proportional to the field strength. The magnetic fields used for imaging can align only a few of every million magnetic nuclei present. However, this is sufficient to produce a useful NMR effect.

Precession and Resonance

When a spinning magnetic nucleus aligns with a magnetic field, it is not fixed; the nuclear magnetic moment precesses, or oscillates, about the axis of the magnetic field, as shown in Figure 3-3. The precessing motion is a physical phenomenon that results from an interaction between the magnetic field and the spinning momentum of the nucleus.

Precession is often observed with a child's spinning top. A spinning top does not stand vertical for long, but begins to wobble, or precess. In this case, the precession is caused by an interaction between the earth's gravitational field and the spinning momentum of the top.

The precession rate (cycles per second) is directly proportional to the strength of the magnetic field. It is this precessing motion

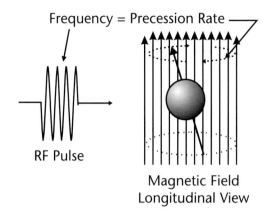

PROTON PRECESSION AND RESONANCE

Frequency = Precession Rate

RF Pulse

Magnetic Field Longitudinal View

Figure 3-3. Magnetic nuclei precession and resonance in a magnetic field.

that makes a nucleus sensitive and receptive to incoming RF energy when the RF frequency matches the precession rate. This precession rate corresponds to the resonant frequency. It is the precessing nuclei, typically protons, that are tuned to receive and transmit RF energy.

Excitation

If a pulse of RF energy with a frequency corresponding to the nuclear precession rate is applied to the material, some of the energy will be absorbed by the individual nuclei. The absorption of energy by a nucleus flips its alignment away from the direction of the magnetic field, as shown in Figure 3-4. This increased energy places the nucleus in an unnatural, or *excited*, state.

In MRI an RF pulse is used that flips some of the nuclei into the transverse plane of the magnetic field. In this excited state the precession is now transformed into a spinning motion of the nucleus around the axis of the magnetic field. It should be noted that this spinning motion is an enhanced precession and is different from the intrinsic spin of a nucleus about its own axis.

The significance of a magnetic nucleus spinning around the axis of the magnetic field is that this motion now generates an RF signal as shown in Figure 3-5. It is this signal, from many nuclei, that is collected to form the MR image.

Relaxation

When a nucleus is in an excited state, it experiences an increased torque from the magnetic field, urging it to realign. The nucleus can return to a position of alignment by transferring its excess energy to other nuclei or the general structure of the material. This process is known as *relaxation*.

Relaxation is not instantaneous following an excitation. It cannot occur until the nucleus is able to transfer its excess energy. How quickly the energy transfer takes place depends on the physical characteristics of the tissue. In fact, the nuclear relaxation rate (or time) is, in many cases, the most significant factor in producing contrast among different types of tissue in an image.

We are more interested in the collective relaxation of many nuclei that produce the magnetization of tissue and will return to this point in the next chapter.

Resonance

The significance of the nuclear precession is that it causes the nucleus to be extremely sensitive, or tuned, to RF energy that has a frequency identical with the precession frequency (rate). This condition is known as *resonance* and is the basis for all MR procedures. NMR is the process in which a nucleus resonates, or "tunes in," when it is in a magnetic field.

Resonance is fundamental to the absorption and emission of energy by many objects and devices. Objects are most effective in exchanging energy at their own resonant frequency. The resonance of an object or device

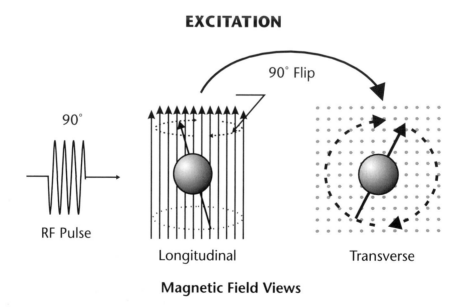

EXCITATION

90°

RF Pulse

90° Flip

Longitudinal

Transverse

Magnetic Field Views

Figure 3-4. The excitation of a magnetic nucleus by the application of a pulse of RF energy.

RF SIGNAL PRODUCTION

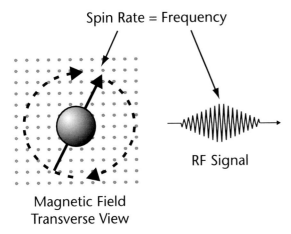

Spin Rate = Frequency

Magnetic Field
Transverse View

RF Signal

Figure 3-5. RF signal production by magnetic nuclei spinning in the transverse plane of a magnetic field.

is determined by certain physical characteristics. Let us consider two common examples.

Radio receivers operate on the principle of resonant frequency. A receiver can select a specific broadcast station because each station transmits a different frequency. Tuning a radio is actually adjusting its resonant frequency. Its receiver is very sensitive to radio signals at its resonant frequency and insensitive to all other frequencies.

The strings of a musical instrument also have specific resonant frequencies. This is the frequency at which the string vibrates to produce a specific audio frequency, or musical note. The resonant frequency of a string depends on the amount of tension. It can be changed, or tuned, by changing the tension. This is somewhat analogous to the resonant frequency of a magnetic nucleus being dependent on the strength of the magnetic field in which it is located.

Larmor Frequency

The resonant frequency of a nucleus is determined by a combination of nuclear characteristics and the strength of the magnetic field. The resonant frequency is also known as the *Larmor frequency*. The specific relationship between resonant frequency and field strength is an inherent characteristic of each nuclide and is generally designated the *gyromagnetic ratio*. The Larmor frequencies [in megahertz (MHz)] for selected nuclides in a magnetic field of 1 T are shown in Table 3-2.

Table 3-2. Larmor Frequencies for Selected Nuclides in a Magnetic Field of 1 T

Nuclide	Larmor Frequency (MHz)
Hydrogen-1	42.58
Fluorine-19	40.05
Phosphorus-31	17.24
Sodium-23	11.26

The fact that different nuclides have different resonant frequencies means that most MR procedures can "look at" only one chemical element (nuclide) at a time.

Field Strength

For all nuclides, the resonant frequency is proportional to the strength of the magnetic field. In a very general sense, increasing the magnetic field strength increases the tension on the nuclei (as with the strings of a musical instrument) and increases the resonant frequency. The fact that a specific nuclide can be tuned to different radio frequencies by varying the field strength (i.e., applying gradients) is used in the imaging process.

Chemical Shift

The resonant frequency of magnetic nuclei, such as protons, is also affected by the structure of the molecule in which they are located.

When a proton, or other magnetic nucleus, is part of a molecule, it is slightly shielded from the large magnetic field. The amount of shielding depends on the chemical composition of the molecule. This means that protons in different chemical compounds will be in slightly different field strengths and will therefore resonate at different frequencies. This change in resonant frequency from one compound to another is known as *chemical shift*. It can be used to perform chemical analysis in the technique of MR spectroscopy and to produce images based on chemical composition. However, in conventional MRI the chemical-shift effect can be the source of an unwanted artifact.

In tissue the chemical shift in resonant frequency between the fat and water is approximately 3.3 ppm, as shown in Figure 3-6. At a field strength of 1.5 T the protons have a basic resonant frequency of approximately 64 MHz. Multiplying this by 3.3 gives a water-fat chemical shift of approximately 210 Hz. At a field strength of 0.5 T the chemical shift would be only 70 Hz.

There are several imaging techniques that can be used to selectively image either the water or fat tissue components. One approach is to suppress either the fat or water signal with specially designed RF pulses. This technique is known as *spectral presaturation* and will be described in Chapter 8. Another technique makes use of the fact that the signals from water and fat are not always in step, or in phase, with each other and can be separated to create either water or fat images.

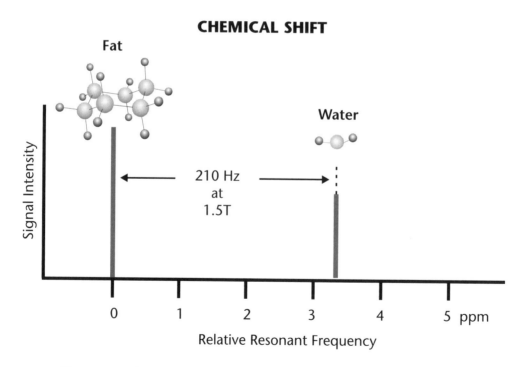

Figure 3-6. The chemical shift effect on the relative resonant frequency of protons in fat and in water.

Mind Map Summary
Nuclear Magnetic Resonance

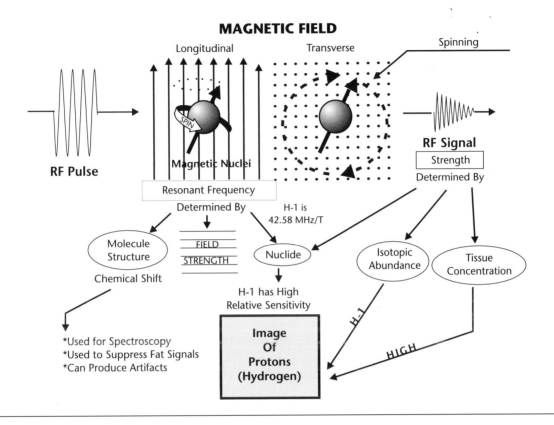

When a magnetic nucleus is located in a strong magnetic field, it resonates. In effect, it becomes a tuned radio receiver and transmitter. The resonance occurs because the spinning nucleus precesses at a rate that is in the radio frequency range. The resonant frequency is determined by three factors. Each specific nuclide has a unique resonant frequency. The resonant frequency is affected to a small degree by the structure of the molecule containing the magnetic nucleus. This, the chemical shift effect, is useful for spectroscopy and to suppress fat signals in images. It can also lead to a certain type of image artifact. The resonant frequency is directly proportional to the strength of the magnetic field. This is useful because it makes it possible to tune the various parts of a body to different frequencies by applying magnetic field gradients.

When an RF pulse is applied to a magnetic nucleus oriented in the longitudinal direction, it can be flipped into the transverse plane. There the nucleus spins around the axis of the magnetic field and generates an RF signal. It is the signals from many spinning nuclei that are collected and used to form the image. It is necessary to have strong signals to produce good images. Signal strength depends on three factors. Each magnetic nuclide has a unique sensitivity or relative signal strength. All chemical elements have several different isotopes, but all isotopes of an element are usually not in the form of magnetic nuclei. Therefore, the abundance of the magnetic isotope for a specific element has a major effect on signal strength. To produce strong signals a tissue must have a relatively high concentration of a chemical element and the most abundant isotope of that element must be magnetic.

Hydrogen is the only chemical element with a high concentration in tissue and body fluids in the form of an isotope that has a magnetic nucleus. Therefore, MR imaging is essentially limited to visualizing only one chemical element, hydrogen.

4

Tissue Magnetization And Relaxation

Introduction And Overview

We have considered the behavior of individual nuclei when placed in a magnetic field. MRI depends on the collective, or net, magnetic effect of a large number of nuclei within a specific voxel of tissue. If a voxel of tissue contains more nuclei aligned in one direction than in other directions, the tissue will be temporarily magnetized in that particular direction. This process is illustrated in Figure 4-1. In the absence of a magnetic field, the nuclei are randomly oriented and produce no net magnetic effect. This is the normal state of tissue before being placed in a magnetic field. When the tissue is placed in a magnetic field, and some of the nuclei align with the field, their combined effect is to magnetize the tissue in the direction of the magnetic field. A large arrow, the *magnetization vector*, is used to indicate the amount and direction of the magnetization. When tissue is placed in a magnetic field, the maximum magnetization that can be produced depends on three factors: (1) the concentration (density) of magnetic nuclei, typically protons, in the tissue voxel; (2) the magnetic sensitivity of the nuclide; and (3) the strength of the magnetic field. Since an imaging magnetic field aligns a very small fraction of the magnetic nuclei, the tissues are never fully magnetized. The amount of tissue magnetization determines the strength of the RF

MAGNETIZATION

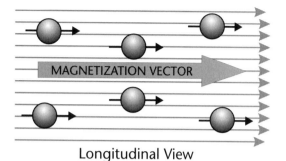

Longitudinal View

Figure 4-1. The magnetization of tissue produced by the alignment of magnetic nuclei (protons) in a magnetic field.

signals emitted by the tissue during an imaging or analytical procedure. This, in turn, affects image quality and imaging time, as explained in Chapter 10.

Let us recall that an MR image is an image of magnetized tissue and that the contrast we see is produced by different levels of magnetization that exist in the different tissues at the time when "the picture is snapped." As we will see in this chapter the level of magnetization at specific times during the imaging process is determined by the three tissue characteristics: proton density (PD), T1, and T2.

We will now see how these characteristics produce image contrast.

Tissue Magnetization

When tissue is placed in a magnetic field, it reaches its maximum magnetization within a few seconds and remains at that level unless it is disturbed by a change in the magnetic field or by pulses of RF energy applied at the resonant frequency. The MRI procedure is a dynamic process in which tissue is cycled through changes in its magnetization during each imaging cycle.

Magnetic Direction

The direction of tissue magnetization is specified in reference to the direction of the applied magnetic field, as shown in Figure 4-2. There are two principle directions that tissue is magnetized during the imaging process. Longitudinal magnetization is when the tissue is magnetized in a direction parallel to the direction of the field. Transverse magnetization is when the direction of tissue magnetization is at a 90° angle with respect to the direction of the magnetic field and is in the transverse plane.

Magnetic Flipping

The direction of tissue magnetization can be changed or flipped by applying a pulse of RF energy. This is done many times throughout the imaging process.

Flip Angle

The angle the magnetization is flipped is determined by the duration and strength of the RF pulse. Pulses are characterized by their *flip angles*.

Pulses with 90° and 180° flip angles are the most common but smaller flip angle pulses are also used in some imaging methods, such as gradient echo imaging.

The 90° Pulse, Saturation and Excitation

When a 90° pulse is applied to longitudinal magnetization, it flips it into the transverse plane as shown in Figure 4-3. This has two effects. First, it reduces the longitudinal magnetization to zero, a condition called *saturation*. It also produces transverse magnetization. As we will soon learn, transverse magnetization is an unstable or *excited* condition. Therefore, when a 90° pulse is applied to longitudinal magnetization, it produces both *saturation* of the longitudinal magnetization and a condition of *excitation* (transverse magnetization).

MAGNETIC DIRECTION

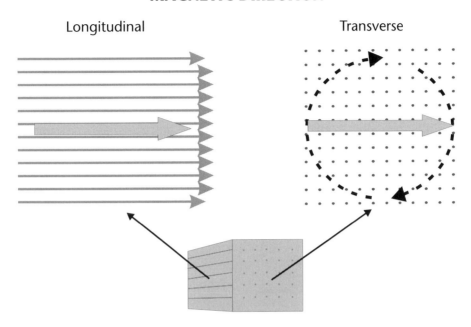

Views of the Magnetic Field in a Voxel

Figure 4-2. Longitudinal and transverse magnetization.

The actual direction of magnetization is not limited to longitudinal or transverse. It can exist in any direction. In principle, magnetization can have both longitudinal and transverse components. Since the two components have distinctly different characteristics, we consider them independently.

Longitudinal Magnetization And Relaxation

As we have seen, when tissue is placed in a magnetic field, it becomes magnetized in the longitudinal direction. It will remain in this state until the magnetic field is changed or until the magnetization is redirected by the application of an RF pulse. If the magnetization is temporarily redirected by an RF pulse, it will then, over a period of time, return to its original longitudinal position. If we consider only the longitudinal magnetization, it regrows after it has been reduced to zero, or saturated. This regrowth, or recovery, of longitudinal magnetization is the *relaxation* process, which occurs after saturation. The time required for the longitudinal magnetization to regrow, or relax, depends on characteristics of the material and the strength of the magnetic field.

Longitudinal magnetization does not grow at a constant rate, but at an exponential rate, as shown in Figure 4-4. An important concept to remember is that the MR image is an image of magnetized tissue with brightness indicating the level of magnetization. During the relaxation process, the level of magnetization is changing. Therefore, the brightness of tissue (if we could see it) is also changing as

MAGNETIC
SATURATION AND EXCITATION

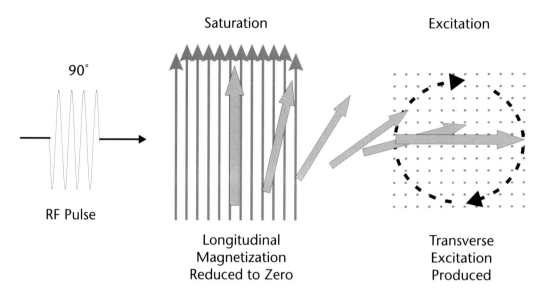

Figure 4-3. The application of a 90° RF pulse to longitudinal magnetization produces saturation of the longitudinal magnetization and creates transverse magnetization, an excited condition.

LONGITUDINAL MAGNETIZATION RELAXATION (GROWTH)

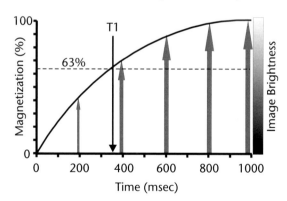

Figure 4-4. The growth of longitudinal magnetization (and tissue brightness) during the relaxation process following saturation.

indicated by the scale on the right of the illustration. Saturation turns the tissue dark and then it recovers brightness during the relaxation period.

The characteristic that varies from one type of tissue to another, and can be used to produce image contrast, is the time required for the magnetization to re-grow, or the relaxation time. Because of its exponential nature, it is difficult to determine exactly when the magnetization has reached its maximum. The convention is to specify the relaxation time in terms of the time required for the magnetization to reach 63% of its maximum. This time, the *longitudinal relaxation time*, is designated T1. The 63% value is used because of mathematical, rather than clinical, considerations. Longitudinal magnetization continues to grow with time, and reaches 87% of its maximum after two T1 intervals, and 95% after three T1 intervals. For practical purposes, the magnetization can be considered fully recovered after approximately three times the T1 value of the specific tissue. We will see later that this must be taken into consideration when setting up an imaging procedure.

T1 Contrast

The time required for a specific level of longitudinal magnetization regrowth varies from tissue to tissue. Figure 4-5 shows the regrowth of two tissues with different T1 values. In this illustration we watch the intensity of brightness of a voxel of tissue during the relaxation process. Let us recall that the brightness of a tissue (RF signal intensity) is determined by the level of magnetization existing in a voxel of tissue at any instant in time. What we see in an image depends on when we "snap the picture" during the relaxation process. The important thing to notice is that the tissue with the shortest T1 has the highest level of magnetization at any particular time. The clinical significance of this is that tissues with short T1 values will be bright in T1-weighted images.

T1 CONTRAST

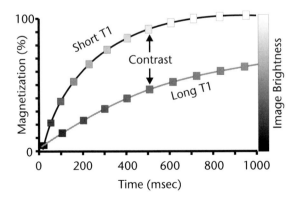

Figure 4-5. The formation of contrast between two tissues with different T1 values.

Table 4-1 lists typical T1 values for various tissues. Two materials establish the lower and upper values for the T1 range: fat has a short T1, and fluid falls at the other extreme (long T1). Therefore, in T1-weighted images, fat is generally bright, and fluid [cerebrospinal fluid (CSF), cyst, etc.] is dark. Most other body tissues are within the range between fat and fluid.

The longitudinal relaxation process involves an interaction between the protons and their immediate molecular environment. The rate of relaxation (T1 value) is related to the naturally occurring molecular motion. The molecular motion is determined by the physical state of the material and the size of the molecules. The relatively rigid structure of solids does not provide an environment for rapid relaxation, which results in long T1 values. Molecular motion in fluids, and fluid-like substances, is more inducible to the relaxation process. In this environment molecular size becomes an important characteristic.

Relaxation is enhanced by a general matching of the proton resonant frequency and the frequency associated with the molecular motions. Therefore, factors that change either of these two frequencies will generally have an effect on T1 values.

Molecular Size

Small molecules, such as water, have faster molecular motions than large molecules, such as lipids. The frequencies associated with the molecular motion of water molecules are both higher and more dispersed over a larger range for the larger molecules. This reduces the match between the frequencies of the protons and the frequencies of the molecular environment. This is why water and similar fluids have relatively long T1 values. Larger molecules, which have slower and less dispersed molecular movement, have a better frequency

Table 4-1. T2 and T1 Values for Various Tissues

Tissue	T2 (msec)	T1 (0.5 T) (msec)	T1 (1.5 T) (msec)
Adipose (Fat)	80	210	260
Liver	42	350	500
Muscle	45	550	870
White Matter	90	500	780
Gray Matter	100	650	920
CSF	160	1800	2400

match with the proton resonant frequencies. This enhances the relaxation process and produces short T1 values. Fat is an excellent example of a large molecular structure that exhibits this characteristic.

Tissues generally contain a combination of water and a variety of larger molecules. Some of the water can be in a relatively free state while other water is bound to some of the larger molecules. In general, the T1 value of the tissue is probably affected by the exchange of water between the free and the bound states. When the water is bound to larger molecular structures, it takes on the motion characteristics of the larger molecule. Factors such as a pathologic process, which alters the water composition of tissue, will generally alter the T1 values.

Magnetic Field Strength Effect

T1 values depend on the strength of the magnetic field. This is because the field strength affects the resonant frequency of the protons. As field strength is increased, the resonant frequency also increases and becomes less matched to the molecular motion frequencies. This results in an increase in T1 values, as indicated in Table 4-1.

Let us now combine two factors to create a T1 image as illustrated in Figure 4-6. One factor is that different tissues have different T1 values and rates of regrowth of longitudinal magnetization. This then causes the different tissues to be at different levels of magnetization (brightness) when the picture is snapped during the relaxation period. Here we see the order of tissue brightness is inversely related to T1 values. In principle, the tissues with short T1 values get brighter faster and are at a higher level when the picture is snapped.

Transverse Magnetization And Relaxation

Transverse magnetization is produced by applying a pulse of RF energy to the magnetized tissue. This is typically done with a 90° pulse, which converts longitudinal magnetization into transverse magnetization. Transverse magnetization is an unstable, or excited, condition and quickly decays after the termination of the excitation pulse. The decay of transverse magnetization is also a relaxation process, which can be characterized by specific relaxation times, or T2 values. Different types of tissue have different T2 values that can be used to discriminate among tissues and contribute to image contrast.

Transverse magnetization is used during the image formation process for two reasons: (1) to develop image contrast based on differences in T2 values; and (2) to generate the RF signals emitted by the tissue. Longitudinal

T1-WEIGHTED IMAGE

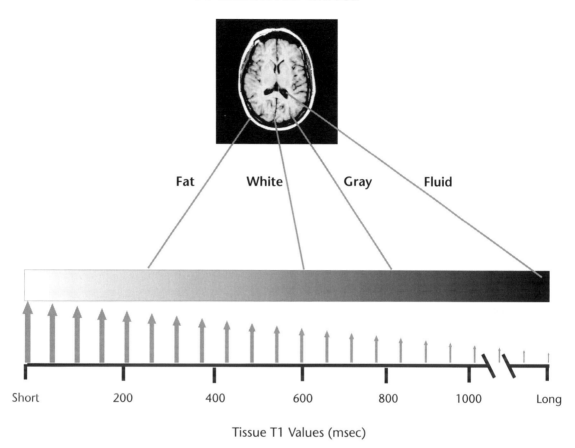

Figure 4-6. A T1 image showing the relationship of tissue brightness (signal intensity) to T1 values and level of magnetization during the longitudinal relaxation process.

magnetization is an RF silent condition and does not produce any signal. However, transverse magnetization is a spinning magnetic condition within each tissue voxel, and that generates an RF signal. As we will see in the next chapter, each imaging cycle must conclude with transverse magnetization to produce the RF signal used to form the image.

The characteristics of transverse magnetization and relaxation are quite different from those for the longitudinal direction. A major difference is that transverse magnetization is an unstable condition and the relaxation process results in the decay, or decrease, in magnetization, as shown in Figure 4-7. The T2 value is the time required for 63% of the initial magnetization to dissipate. After one T2, 37% of the initial magnetization is present.

T2 Contrast

The difference in T2 values of tissues is the source of contrast in T2-weighted images. This is illustrated in Figure 4-8. Here we watch two tissues, with different T2 values, during the relaxation process. We see that they are both getting darker with time as the magnetization

---

TRANSVERSE MAGNETIZATION RELAXATION (DECAY)

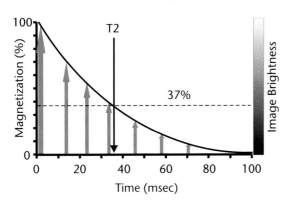

Figure 4-7. The decay of transverse magnetization during the relaxation process and the associated tissue brightness.

T2 CONTRAST

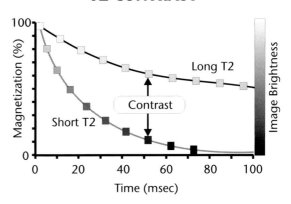

Figure 4-8. The formation of T2 contrast during the decay of transverse magnetization.

decays. However, they are not getting darker at the same rate. The tissue with the shorter T2 becomes darker faster leaving the tissue with the longer T2 to be bright at times during the relaxation time.

What we will actually see in a T2-weighted image, as shown in Figure 4-9, depends on the level of magnetization at the time when we snap the picture. The important thing to observe here is that the tissues with *long* T2 values are bright in T2 images.

In general, a T2-weighted image appears to be a reversal of a T1-weighted image. Tissues that are bright in one image are dark in the other image. This is because of a combination of two factors. One factor is that T1 and T2 values are generally related. Even though T2 values are much shorter than T1 values, as shown in Table 4-1, they are somewhat proportional. Tissues with long T1 values usually have long T2 values. The other factor is that the order of brightness in a T2 image is in the same direction as the T2 values. Remember, it was a reversed relationship for T1 images.

The decay of transverse magnetization (i.e., relaxation) occurs because of a dephasing among individual nuclei (protons) within the individual voxels, as shown in Figure 4-10.

Two basic conditions are required for transverse magnetization: (1) the magnetic moments of the nuclei must be oriented in the transverse direction, or plane; and (2) a majority of the magnetic moments must be in the same direction, or in phase, within the transverse plane. When a nucleus has a transverse orientation, it is actually spinning around an axis that is parallel to the magnetic field.

After the application of a 90° pulse, the nuclei have a transverse orientation and are rotating together, or in phase, around the magnetic field axis. This rotation or spin is a result of the normal precession discussed earlier. The precession rate, or resonant frequency, depends on the strength of the magnetic field where the nuclei are located. Nuclei located in field areas with different strengths spin (precess) at different rates. Even within a very small volume of tissue, nuclei are in slightly different magnetic field strengths. As a result, some nuclei spin faster than others. Also, there are interactions (spin-spin interactions) among the spinning nuclei. After a short period of time, the nuclei are not spinning in phase. As the directions of the nuclei begin to spread and they dephase, the magnetization of the tissue decreases. A short time later, the nuclei are randomly oriented in the transverse plane, and there is no transverse magnetization.

Proton Dephasing

Two major effects contribute to the dephasing of the nuclei and the resulting transverse relaxation. In the imaging process the spin echo technique is used to separate the two sources of dephasing, as we will see in Chapter 6.

T2 Tissue Characteristics

One effect is the exchange of energy among the spinning nuclei (spin-spin interactions), which results in relatively slow dephasing and loss of magnetization. The rate at which this occurs is determined by characteristics of the tissue. It is this dephasing activity that is characterized by the T2 values as shown in Table 4-1.

T2* Magnetic Field Effects

A second effect, which produces relatively rapid dephasing of the nuclei and loss of transverse magnetization, is the inherent inhomogeneity of the magnetic field within

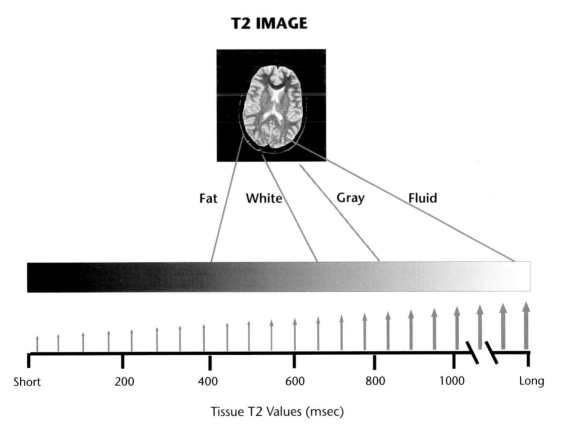

Figure 4-9. A T2 image showing the relationship of tissue brightness (signal intensity) to T2 values.

TRANSVERSE MAGNETIZATION DECAY

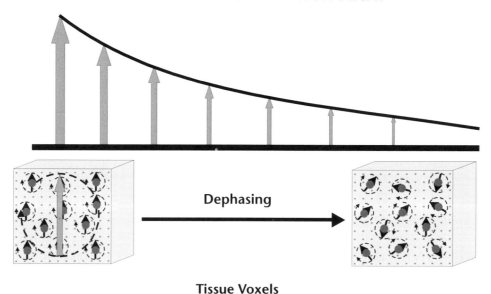

Figure 4-10. The dephasing of protons that produces transverse magnetization decay.

each individual voxel. The field inhomogeneities are sufficient to produce rapid dephasing. This effect, which is different from the basic T2 characteristics of the tissue, tends to mask the true relaxation characteristics of the tissue. In other words, the actual transverse magnetization relaxes much faster than the tissue characteristics would indicate. This real relaxation time is designated as T2*. The value of T2* is usually much less than the tissue T2 value, as illustrated in Figure 4-11. Several factors can contribute to field inhomogeneities and to T2* decay. One is the general condition of the magnetic field. Some fields are more homogeneous than others. Another factor is that different tissues or materials in the body might have different magnetic *susceptibilities*. *Susceptibility* is a characteristic of a material that determines its ability to become magnetized when it is in a magnetic field. If a region of tissue contains materials with different susceptibilities, this results in a reduction of field homogeneity.

Magnetic Susceptibility

The magnetization of tissue that we have been discussing is a *nuclear* magnetic effect produced by the alignment of magnetic nuclei in a magnetic field. Other materials can become magnetized by other, non-nuclear effects.

Many materials are *susceptible* to magnetic fields and become magnetized when located in fields. The susceptibility of a material is determined by the orbital electrons in the atom rather than the magnetic properties of the nucleus. Significant susceptibility is present only when there are unpaired electrons in the outer orbit.

There are three general types of materials with respect to magnetic susceptibility: *diamagnetic, paramagnetic*, and *ferromagnetic*. The primary characteristic of each type is the amount and direction of magnetization that the material develops when placed in a magnetic field. There are situations when each type plays a role in the MR imaging process.

FACTORS AFFECTING TRANSVERSE MAGNETIZATION RELAXATION

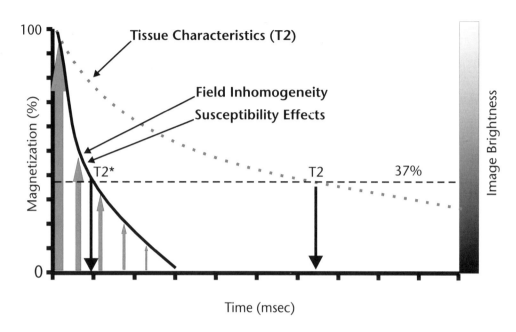

Figure 4-11. Comparison of relaxation produced by the T2 characteristics of tissue and the T2* effects associated with magnetic field inhomogeneities.

Contrast Agents

The inherent tissue characteristics (PD, T1, and T2) do not always produce adequate contrast for some clinical objectives. It is possible to administer materials (i.e., contrast agents) that will alter the magnetic characteristics within specific tissues or anatomical regions. There are several different types of contrast agents, which will now be considered. Contrast agents used in MRI are generally based on relaxation effects.

Diamagnetic Materials

Diamagnetic materials have negative and relatively low magnetic susceptibility. This means that they develop only low levels of magnetization and it is in a direction opposite to the direction of the magnetic field. Although many biological molecules are diamagnetic, this is not a significant factor in MR imaging.

Paramagnetic Materials

Paramagnetic materials play an important role in contrast enhancement. They are materials with unpaired electrons that give each atom a permanent magnetic property. In paramagnetic materials each atom is magnetically independent, which distinguishes it from other materials to be discussed later.

Paramagnetic substances include metal ions such as gadolinium, manganese, iron, and chromium. Other substances such as nitroxide free radicals and molecular oxygen also have paramagnetic properties.

Gadolinium has seven unpaired electrons in its orbit, which give it a very strong magnetic property. It must be chelated to reduce its toxicity. An example is gadolinium chelated to diethylene triamine penta-acetic acid (GaDTPA).

When a paramagnetic substance, such as gadolinium, enters an aqueous solution, it affects the relaxation rate of the existing protons. It does not produce a signal itself. In relatively low concentrations, the primary effect is to increase the rate of longitudinal relaxation and shorten the value of T1. In principle, the fluctuating magnetic field from the individual paramagnetic molecules enhances the relaxation rate. The primary result is an increase in signal intensity with T1-weighted images. It is classified as a positive contrast agent.

Signal intensity will generally increase with the concentration of the paramagnetic agents until a maximum intensity is reached. This intensity is very dependent on the imaging parameters. Higher concentrations will generally produce a reduction of signal intensity. This occurs because the transverse relaxation rate is also increased, which results in a shortening of the T2 value.

Superparamagnetic Materials

When materials with unpaired electrons are contained in a crystalline structure, they produce a stronger magnetic effect (susceptibility) in comparison with the independent molecules of a paramagnetic substance. The susceptibility of superparamagnetic materials is several orders of magnitude greater than that of paramagnetic materials. These materials are in the form of small particles. Iron oxide particles are an example.

The particles produce inhomogeneities in the magnetic field, which results in rapid dephasing of the protons in the transverse plane and a shortening of T2.

Superparamagnetic materials in the form of large particles generally reduce signal intensity and are classified as negative contrast agents. When in the form of very small particles, they reduce T1 and increase signal intensity.

Ferromagnetic Materials

Ferromagnetic is the name applied to iron and only a few other materials that have magnetic properties like iron. These materials have a very high susceptibility and develop a high level of magnetism when placed in a magnetic field.

Mind Map Summary
Tissue Magnetization And Relaxation

TISSUE MAGNETIZATION

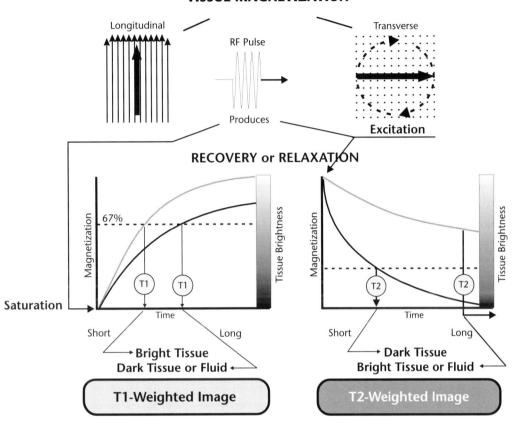

When tissue containing magnetic nuclei, i.e., protons, is placed in a strong magnetic field, the tissue becomes magnetized. It is initially magnetized in the longitudinal direction. However, by applying a pulse of RF energy the magnetization can be flipped into the transverse plane. Both longitudinal and transverse magnetization have characteristics that can be used to develop image contrast. An imaging procedure can be adjusted to display the different types of contrasts.

When a 90° RF pulse is applied to longitudinal magnetization, it produces two effects. First, it temporarily destroys the longitudinal magnetization, a condition known as *saturation*. It also produces transverse magnetization, a condition known as *excitation* because transverse magnetization is an unstable excited state.

After a saturation pulse is applied, the longitudinal magnetization will recover or regrow, a process known as *relaxation*. The rate of regrowth is a characteristic of each specific tissue and is described by its T1 value, the longitudinal relaxation time. A tissue with a short T1 will recover its magnetization fast and will appear bright in a T1-weighted image. Tissues with longer T1 values will recover magnetization somewhat slower and will be relatively dark in T1-weighted images.

Following the production of transverse magnetization by the RF pulse the magnetization begins to decay or relax. The rate of relaxation is a characteristic of each specific tissue and is expressed by the T2 values, the transverse relaxation time. A tissue with a short T2 will lose its transverse magnetization rapidly and will appear relatively dark in T2-weighted images. Tissues and body fluids with long T2 values will retain their transverse magnetization longer and will appear bright in T2-weighted images.

5

The Imaging Process

Introduction And Overview

The MR imaging process consists of two major functions as shown in Figure 5-1. The first is the *acquisition* of RF signals from the patient's body and the second is the mathematical *reconstruction* of an image from the acquired signals.

In this chapter we will develop a general overview of the imaging process and set the stage for considering the different methods and techniques that are used to produce optimum images for various clinical needs.

k Space

During the acquisition process the signals are collected, digitized, and stored in computer memory in a configuration known as *k space*.

The k space is divided into lines of data that are filled one at a time. One of the general requirements is that the k space must be completely filled before the image reconstruction can be completed. The size of k space (number of lines) is determined by the requirements for image detail and will be discussed in Chapters 9 thru 11.

Acquisition

The acquisition process consists of an imaging cycle that is repeated many times. The time required for a complete acquisition is determined by the duration of the cycle multiplied by the number of cycles. The duration of a cycle is TR (Time of Repetition), the adjustable protocol factor that is used to select the different

THE MR IMAGING PROCESS

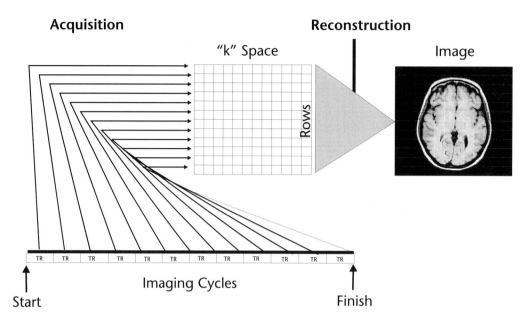

Figure 5-1. The two functions, *acquisition* and *reconstruction*, that make up the MR image production process.

types of image contrast. Also, the number of cycles used in an acquisition is adjustable. The number of cycles depends on the quality of the image that is required. The complete relationship between number of imaging cycles and image quality characteristics is described in Chapter 10.

Reconstruction

The image reconstruction process is usually fast compared to the acquisition process and generally does not require any decisions or adjustments by the operator.

Imaging Protocol

Each imaging procedure is controlled by a protocol that has been entered into the computer. Issues that must be considered in selecting, modifying, or developing a protocol for a specific clinical procedure include:

- The imaging method to be used
- The image types (PD, T1, T2, etc.)
- Spatial characteristics (slice thickness, number, etc.)
- Detail and visual noise requirements
- Use of selective signal suppression techniques
- Use of artifact reduction techniques

In the following chapters we will address each of these issues and the specific protocol factors that are used to produce the desired image characteristics.

Imaging Methods

There are several different imaging methods that can be used to create MR images. The principal difference among these methods is the sequence in which the RF pulses and gradients are applied during the acquisition process.

Therefore, the different methods are often referred to as the different *pulse sequences*. An overview of the most common methods is shown in Figure 5-2. As we see, the different methods are organized in a hierarchy structure. For each imaging method there is a set of factors that must be adjusted by the user to produce specific image characteristics.

The selection of a specific imaging method and factor values is generally based on requirements for contrast sensitivity to a specific tissue characteristic (PD, T1, T2) and acquisition speed. However, other characteristics such as visual noise and the sensitivity to specific artifacts might vary from method to method.

All of the imaging methods belong to one or both of the two major families, *spin echo* or *gradient echo*. The difference between the two

families of methods is the process that is used to create the echo event at the end of each imaging cycle. For the spin echo methods, the echo event is produced by the application of a 180° RF pulse, as will be described in Chapter 6. For the gradient echo methods the event is produced by applying a magnetic field gradient, as described in Chapter 7. Each method has very specific characteristics and applications.

The Imaging Cycle

A common characteristic of all methods is that there are two distinct phases of the image acquisition cycle, as shown in Figure 5-3. One phase is associated with longitudinal magnetization and the other with transverse magnetization. In general, T1 contrast is developed during the longitudinal magnetization phase and T2 contrast is

IMAGING METHODS

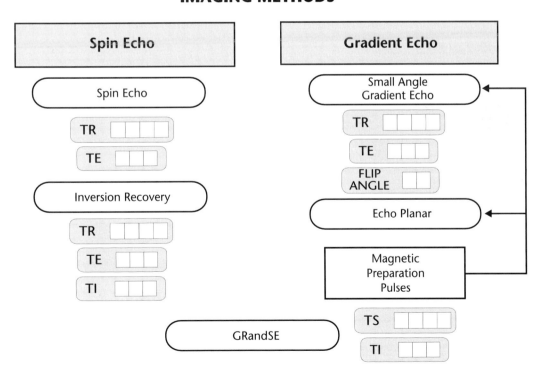

Figure 5-2. The principal spin echo and gradient echo imaging methods. GRandSE, or GRASE, is a combination of the two methods.

developed during the transverse magnetization phase. PD contrast is always present, but becomes most visible when it is not overshadowed by either T1 or T2 contrast. The predominant type of contrast that ultimately appears in the image is determined by the duration of the two phases and the transfer of contrast from the longitudinal phase to the transverse phase.

The duration of the two phases (longitudinal and transverse) is determined by the selected values of the protocol factors, TR (Time of Repetition) and TE (Time to Echo).

TR

TR is the time interval between the beginning of the longitudinal relaxation, following saturation, and the time at which the longitudinal magnetization is converted to transverse magnetization by the excitation pulse. This is when the picture is snapped relative to the longitudinal magnetization.

Because the longitudinal relaxation takes a relatively long time, TR is also the duration of the image acquisition cycle or the cycle repetition time (Time of Repetition).

TE

TE is the time interval between the beginning of transverse relaxation following excitation and when the magnetization is measured to produce image contrast. This happens at the *echo event* and is when the picture is snapped relative to the transverse magnetization. Therefore, TE is the Time to Echo event.

ONE IMAGING CYCLE

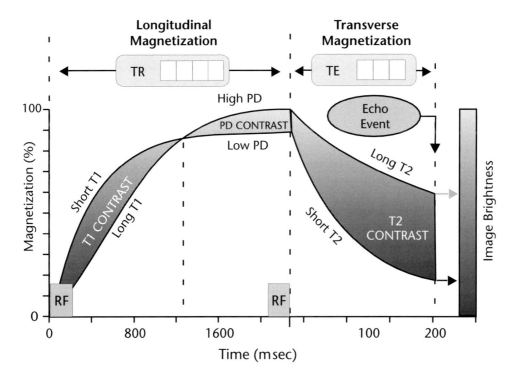

Figure 5-3. The longitudinal and transverse magnetization phases of an imaging cycle. T1 and PD contrast are produced during the longitudinal phase and T2 contrast is produced during the transverse phase.

Magnetic Resonance Imaging

Excitation

The transition from the longitudinal magnetization phase to the transverse magnetization phase is produced by applying an RF pulse. This is generally known as the excitation process because the transverse magnetization represents a more unstable or "excited" state than longitudinal magnetization.

The excitation pulse is characterized by a flip angle. A 90° excitation pulse converts all of the existing longitudinal magnetization into transverse magnetization. This type of pulse is used in the spin echo methods. However, there are methods that use excitation pulses with flip angles that are less than 90°. Small flip angles (<90°) convert only a fraction of the existing longitudinal magnetization into transverse magnetization and are used primarily to reduce acquisition time with the gradient echo methods described in Chapter 7.

The Echo Event and Signals

The transverse magnetization phase terminates with the echo event, which produces the RF signal. This is the signal that is emitted by the tissue and used to form the image. The echo event is produced by applying either an RF pulse or a gradient pulse to the tissue, as will be described in Chapters 6 and 7.

Contrast Sensitivity

In MRI the usual procedure is to select one of the tissue characteristics (PD, T1, T2) and then adjust the imaging process so that it has maximum, or at least adequate, contrast sensitivity for that specific characteristic. This produces an image that is heavily *weighted* by that characteristic. The contrast sensitivity of the imaging process and the resulting image contrast is determined by the specific imaging method and the combination of imaging

protocol factor values, which we will consider in much more detail in later chapters. The discussion in this chapter will be based on the conventional spin echo method that uses only two factors, TR and TE, to control contrast sensitivity. However, it establishes some principles that apply to all methods.

T1 Contrast

During the relaxation (regrowth) of longitudinal magnetization, different tissues will have different levels of magnetization because of their different growth rates, or T1 values. Figure 5-4 compares two tissues with different T1 values.

The tissue with the shorter T1 value experiences a faster regrowth of longitudinal magnetization. Therefore, during this period of time it will have a higher level of magnetization, produce a more intense signal, and appear brighter in the image. In T1-weighted images brightness or high signal intensity is associated with short T1 values.

At the beginning of each imaging cycle, the longitudinal magnetization is reduced to zero (saturation) by an RF pulse, and then allowed to regrow, or relax. This is what happens in the spin echo method. In some other imaging methods, as we will see in the next two chapters, the cycle might begin with either partially saturated or inverted longitudinal magnetization. In all cases, T1 contrast is formed during the regrowth process. At a time determined by the selected TR value, the cycle is terminated and the magnetization value is converted to transverse, measured and displayed as a pixel intensity, or brightness, and a T1-weighted image is produced.

In principle, at the beginning of each imaging cycle all tissues are dark. As the tissues regain longitudinal magnetization, they become brighter. The brightness, or intensity, with which they appear in the image depends on

T1 CONTRAST
Longitudinal Phase

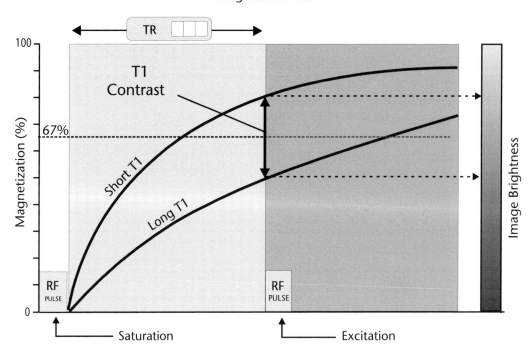

Figure 5-4. The amount of T1 contrast captured during the longitudinal magnetization phase is determined by the value of TR that is selected by the operator.

when during the regrowth process the cycle is terminated and the picture is snapped. This is determined by the selected TR value. When a short TR is used, the regrowth of the longitudinal magnetization is interrupted before it reaches its maximum. This reduces signal intensity and tissue brightness within the image but produces T1 contrast.

Increasing TR increases signal intensity and brightness up to the point at which magnetization is fully recovered, which is determined by the PD of each tissue. For practical purposes, this occurs when the TR exceeds approximately three times the T1 value for the specific tissues. Although it takes many cycles to form a complete image, the longitudinal magnetization is always measured at the same time in each cycle as determined by the setting of TR.

To produce a T1-weighted image, a value for TR must be selected to correspond with the time at which T1 contrast is significant between the two tissues. Several factors must be considered in selecting TR. If T1 contrast is represented by the ratio of the tissue magnetization levels, it is at its maximum very early in the relaxation process. However, the low magnetization levels present at that time do not generally produce adequate RF signal levels for many clinical applications. The selection of a longer TR produces greater signal strength but less T1 contrast.

The selection of TR must be appropriate for the T1 values of the tissues being imaged. If a TR value is selected that is equal to the T1 value of a tissue, the picture will be snapped when the tissue has regained 63% of its magnetization. This represents the time when there is maximum contrast between tissues with small differences in T1 values.

Proton Density (PD) Contrast

The density, or concentration, of protons in each tissue voxel determines the maximum level of magnetization that can be obtained. Differences in PD among tissues can be used to produce image contrast, as illustrated in Figure 5-5. Here we see the growth of longitudinal magnetization for two tissues with the same T1 values but different relative PDs. The tissue with the lowest PD (80) reaches a maximum magnetization level that is only 80% that of the other tissue. The difference in magnetization levels at any point in time is because

of the difference in PD and is therefore the source of PD contrast.

Although there is some PD contrast early in the cycle, it is generally quite small in comparison to the T1 contrast.

The basic difference between T1 contrast and PD contrast is that T1 contrast is produced by the rate of growth (relaxation), and PD contrast is produced by the maximum level to which the magnetization grows. In general, T1 contrast predominates in the early part of the relaxation phase, and PD contrast predominates in the later portion. T1 contrast gradually gives way to PD contrast as magnetization approaches the maximum value. A PD-weighted image is produced by selecting a relatively long

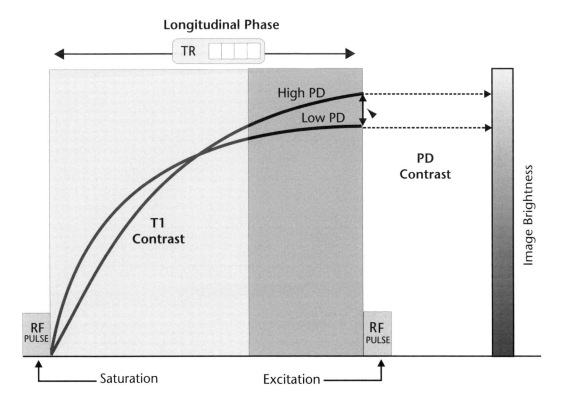

Figure 5-5. Proton density (PD) contrast is captured by setting TR to relatively long values. At that time the magnetization is determined by PD and is not T1 as in the earlier part of the cycle.

TR value so that the image is created or "the picture is snapped" in the later portion of the relaxation phase, where tissue magnetizations approach their maximum values. The TR values at which this occurs depend on the T1 values of the tissues being imaged.

It was shown earlier that tissue reaches 95% of its magnetization in three T1s. Therefore, a TR value that is at least three times the T1 values for the tissues being imaged produces almost pure PD contrast.

T2 Contrast

Now let us turn our attention to the transverse phase. During the decay of transverse magnetization, different tissues will have different levels of magnetization because of different decay rates, or T2 values. As shown in Figure 5-6, tissue with a relatively long T2 value will have a higher level of magnetization, produce a more intense signal, and appear brighter in the image than a tissue with a shorter T2 value.

Figure 5-6 shows the decay of transverse magnetization for tissues with different T2 values. The tissue with the shortest T2 value loses its magnetization faster than the other tissues.

The difference in T2 values of tissue can be translated into image contrast. For the purpose of this illustration we assume that the two tissues begin their transverse relaxation with the levels of magnetization determined by the PD. This is the usual case where the PD contrast

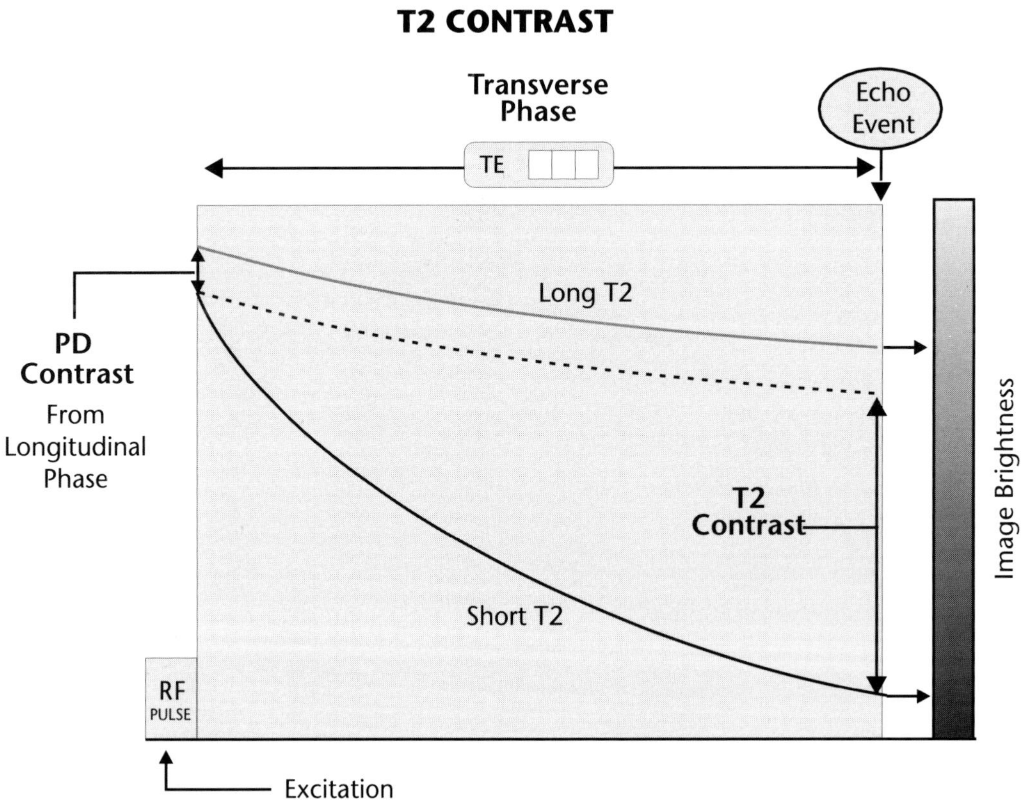

T2 CONTRAST

Figure 5-6. The formation of T2 contrast during the transverse magnetization phase. The amount of T2 contrast captured depends on the selected value of TE, the Time to Echo event.

present at the end of the longitudinal phase carries over to the beginning of the transverse phase. In effect, the transverse phase begins with PD contrast but adds T2 contrast as time elapses. The decay of the magnetization proceeds at different rates because of the different T2 values. The tissue with the longer T2 value maintains a higher level of magnetization than the other tissue and will remain bright longer. The difference in the tissue magnetizations at any point in time represents contrast.

At the beginning of the cycle there is no T2 contrast, but it develops and increases throughout the relaxation process. At the echo event the magnetization levels are converted into RF signals that are displayed as image pixel brightness; this is the time to echo event (TE) and is selected by the operator. Maximum T2 contrast is generally obtained by using a relatively long TE. However, when a very long TE value is used,

the magnetization and the RF signals might be too low to form a useful image. In selecting TE values, a compromise must often be made between T2 contrast and good signal intensity.

The transverse magnetization characteristics of tissue (T2 values) are, in principle, added to the longitudinal characteristics carried over from the longitudinal phase (e.g., T1 and PD) to form the MR image. Usually we do not want to add T2 contrast to T1 contrast. That is because these two types of contrast oppose each other. Remember in Chapter 1 we saw that tissues that are bright in T1 images are dark in T2 images. This means that if we were to mix T1 and T2 contrast in the same image, one would cancel the other. When setting up a protocol for a T2 image it is necessary to use a long TR (in addition to a long TE) so that no, or very little, T1 contrast carries over to the transverse phase.

Mind Map Summary
The Imaging Process

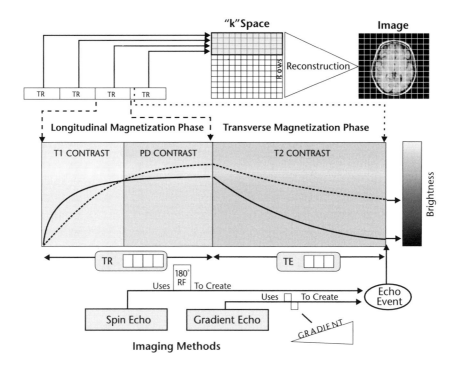

Imaging Methods

The MR imaging process is one of creating contrast among tissues based on their magnetic characteristics. The primary characteristics are proton density (PD), T1, and T2. It is a dynamic activity in which the magnetization levels of the various tissues are undergoing almost constant change. During each imaging cycle there are two distinct magnetization phases: longitudinal and transverse. Different types of contrast are developed in each of these phases.

After application of a saturation pulse, which reduces the longitudinal magnetization to zero, the magnetization begins to regrow, a process known as relaxation. The rate of regrowth for a specific tissue is determined by that tissue's T1 value. Tissues with short T1 values grow faster than tissues with long T1 values. During this regrowth, T1 contrast in the form of different levels of magnetization is created among the tissues. This is the contrast that will be displayed in an image if the protocol parameters are set to produce a T1-weighted image. In a T1-weighted image, tissues with short T1 values will be bright. Tissues and fluid with long T1 values will be darker.

When an RF pulse is applied to longitudinal magnetization, it converts (flips) it to transverse magnetization, an unstable excited magnetic condition that decays with time. This decay process is the transverse magnetization relaxation process. The rate of decay of a specific tissue depends on that tissue's T2 value. Tissues with short T2 values decay faster than tissues with longer T2 values. When the imaging protocol factors are set to produce a T2-weighted image, tissues with short T2 values will be dark and tissues and fluids with longer T2 values will be bright.

6

Spin Echo Imaging Methods

Introduction And Overview

Spin echo is the name of the process that uses an RF pulse to produce the echo event. It is also the name for one of the specific imaging methods within the spin echo family of imaging methods; all of which use the spin echo process. We will first discuss the spin echo process and see how an RF pulse can produce an echo event and signal and then consider the spin echo methods.

The Spin Echo Process

The decay of transverse magnetization (i.e., relaxation) occurs because of dephasing among individual nuclei, as described in Chapter 4.

Let us recall that two basic conditions are required for transverse magnetization: (1) the magnetic moments of the nuclei must be oriented in the transverse direction, or plane, and (2) a majority of the moments must be in the same direction within the transverse plane. When a nucleus has a transverse orientation, it is actually precessing or rotating around an axis that is parallel to the magnetic field.

After the application of a 90° excitation pulse, the nuclei have a transverse orientation and are precessing together, or in-phase, around the magnetic field axis. This is the normal precession discussed earlier but flipped into the transverse plane. However, within an individual voxel some nuclei precess or spin faster than others. After a short period of time, the nuclei are not spinning in-phase. As the directions of the nuclei begin to spread, the magnetization of the tissue decreases. A short time later, the nuclei are randomly oriented in the transverse plane; there is no transverse magnetization.

The two factors that contribute to the dephasing of the nuclei and the resulting transverse relaxation will now be reviewed again here. One is an exchange among the spinning nuclei (spin-spin interactions), which results in relatively slow dephasing and loss of magnetization. The rate at which this occurs is determined by characteristics of the tissue. It is this dephasing activity that is characterized by the T2 values and the source of contrast that we want to capture in T2 images. A second factor, which produces relatively rapid dephasing of the nuclei and loss of transverse magnetization, is the inhomogeneity of the magnetic field. Even within a small volume of tissue, the field inhomogeneities are sufficient to produce rapid dephasing. This effect, which is generally unrelated to the T2 characteristics of the tissue, tends to mask the true relaxation characteristics of the tissue. In other words, the actual transverse magnetization relaxes much faster than the tissue characteristics would indicate. We remember that this real relaxation time is designated as T2*. The value of T2* is always much less than the tissue T2 value. As a result, the transverse magnetization disappears before T2 contrast can be formed.

We are about to discover that spin echo is a process for recovering the lost transverse magnetization and making it possible to produce images of the three tissue characteristics, including T2.

An RF signal is produced whenever there is transverse magnetization. Immediately after an excitation pulse, a so-called free induction decay (FID) signal is produced. The intensity of this signal is proportional to the level of transverse magnetization. Both decay rather rapidly because of the magnetic field inhomogeneities just described. The FID signal is not used in the spin echo methods. It is used in the gradient echo methods to be described in Chapter 7.

The spin echo process is used to compensate for the dephasing and rapid relaxation caused by the field inhomogeneities and to restore the magnetization to the level that depends only on the tissue T2 characteristics. The sequence of events in the spin echo process is illustrated in Figure 6-1.

Transverse magnetization is produced with a 90° RF excitation pulse that flips the longitudinal magnetization into the transverse plane. Immediately following the RF pulse, each voxel is magnetized in the transverse direction. However, because of the local magnetic field inhomogeneities within each voxel, the protons precess at different rates and quickly slip out of phase. This produces the rapid decay characterized by T2* and the associated FID signal. At this time the protons are still rotating in the transverse plane, but they are out of phase.

If a 180° pulse is applied to the tissue containing these protons, it flips the protons around an axis in the transverse plane; this reverses their direction of rotation as illustrated in Figure 6-2. This causes the fast protons to be located behind the slower ones. As the faster protons begin to catch up with the slower ones, they regain a common alignment, or come back into phase. This, in turn, causes the transverse magnetization to reappear and form the echo event. However, the magnetization does not grow to the initial value because the relaxation (dephasing) produced by the tissue is not reversible. The rephasing of the protons causes the magnetization to build up to a level determined by the T2 characteristics of the tissue. As soon as the magnetization reaches this

THE SPIN ECHO PROCESS

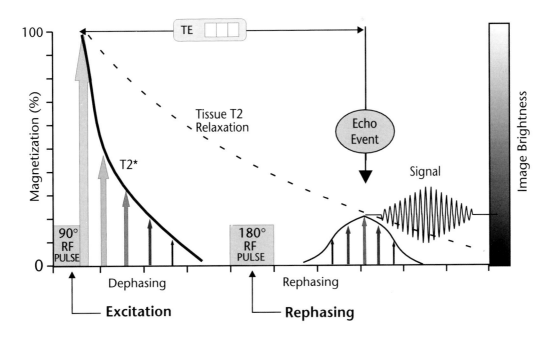

Figure 6-1. The spin echo process showing the use of a 180° pulse to rephase the protons and to produce an echo event.

maximum, the protons begin to move out of phase again, and the transverse magnetization dissipates. Another 180° pulse can be used to produce another rephasing. In fact, this is what is done in multi-echo imaging and will be described later in this chapter.

RF Pulse Sequence

The different imaging methods are produced by the type (flip angle) and time intervals between the applied RF pulses. The basic pulse sequence for the spin echo method is shown in Figure 6-3. Each cycle begins with a 90° excitation pulse that produces the initial transverse magnetization and a later 180° pulse that rephases the protons to produce the echo event.

The time between the initial excitation and the echo signal is TE. This is controlled by

adjusting the time interval between the 90° and the 180° pulses, which is 1/2 TE.

The Spin Echo Method

This method can be used to produce images of the three basic tissue characteristics: PD, T1, and T2. The sensitivity to a specific characteristic is determined by the values selected for the two time intervals or imaging factors, TR and TE.

The process of creating images with the three types of contrast (PD, T1, and T2) described in the last chapter was a description of the spin echo method. There we saw that the type of image that was produced depended on the values selected for the two protocol factors, TR and TE. We will now review that process with a few more details specifically as it applies to the spin echo method.

THE SPIN ECHO PROCESS

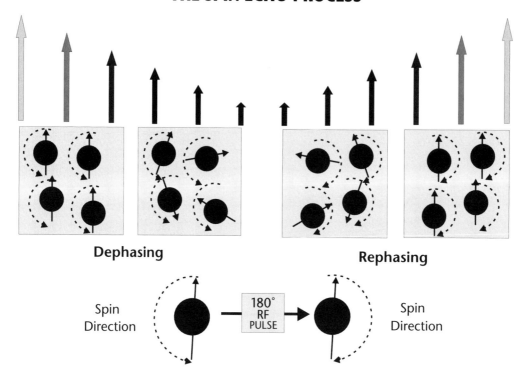

Figure 6-2. The 180° pulse sets up the protons so that they rephase.

Proton Density (PD) Contrast

PD contrast develops as the longitudinal magnetization approaches its maximum, which is determined by the PD of each specific tissue. Therefore, relatively long TR values are required to produce a PD-weighted image. Short TE values are generally used to reduce T2 contrast contamination and to maintain a relatively high signal intensity.

T1 Contrast

To produce image contrast based on T1 differences between tissues, two factors must be considered. Since T1 contrast develops during the early growth phase of longitudinal magnetization, relatively short TR values must be used to capture the contrast. The second factor is to preserve the T1 contrast during the time of transverse relaxation. The basic problem is that if T2 contrast is allowed to develop, it generally counteracts T1 contrast. This is because tissues with short T1 values usually have short T2 values. The problem arises because tissues with short T1s are generally bright, whereas tissues with short T2s have reduced brightness when T2 contrast is present. T2 contrast develops during the TE time interval. Therefore, a T1-weighted image is produced by using short TR values and short TE values.

SPIN ECHO TIME INTERVALS

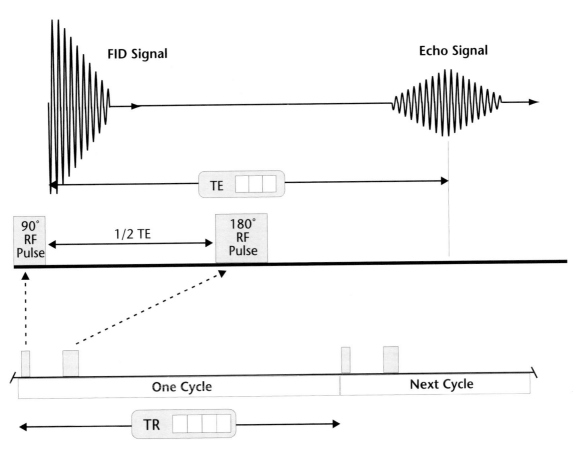

Figure 6-3. The RF pulses and time intervals in a spin echo imaging cycle.

T2 Contrast

The first step in producing an image with significant T2 contrast is to select a relatively long TR value. This minimizes T1 contrast contamination and the transverse relaxation process begins at a relatively high level of magnetization. Long TE values are then used to allow T2 contrast time to develop.

The spin echo method is the only method that produces true T2 contrast. That is because it is able to rephase the protons and remove the T2* effect.

Multiple Spin Echo

It is possible to produce a series of echo events within one cycle as illustrated in Figure 6-4. This is done by applying several 180° pulses after each 90° excitation pulse. The advantage is that echo events with different TE values are produced in one acquisition cycle. Separate images are formed for each TE value. This makes it possible to create both a PD image (short TE) and a T2 image (long T2) in the same acquisition.

Table 6-1 summarizes the combination of TR and TE values used to produce the three

MULTIPLE ECHO IMAGING

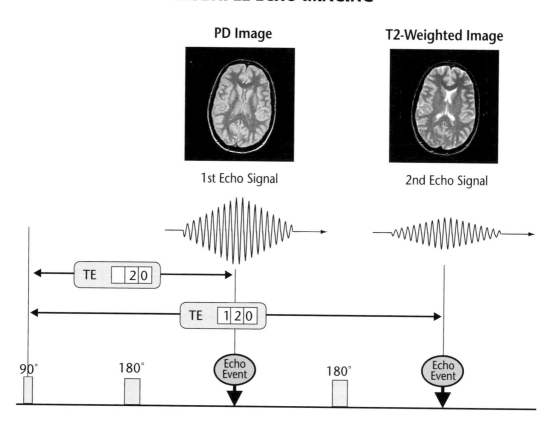

Figure 6-4. A multiple spin echo imaging that produces both a PD and T2 image in the same acquisition.

basic image types with the spin echo method. Optimum values of TR and TE for a specific protocol might vary because of considerations for other factors such as image acquisition time, number of slices, etc.

Inversion Recovery

Inversion recovery is a spin echo imaging method used for several specific purposes. One application is to produce a high level of T1 contrast and a second application is to suppress the signals and resulting brightness of fat and fluids. The inversion recovery pulse sequence is obtained by adding an additional 180° pulse to

the conventional spin echo sequence, as shown in Figure 6-5. The pulse is added at the beginning of each cycle where it is applied to the longitudinal magnetization carried over from the previous cycle. Each cycle begins as the 180° pulse inverts the direction of the longitudinal magnetization. The regrowth (recovery) of the magnetization starts from a negative (inverted) value, rather than from zero, as in the spin echo method.

The inversion recovery method, like the spin echo method, uses a 90° excitation pulse to produce transverse magnetization and a final 180° pulse to produce a spin echo signal. That is

Table 6-1. Selection of TR and TE values to produce the three image types with spin echo method. Values shown are typical but can be varied to some extent to accommodate specific imaging conditions.

	T1 Image	PD Image	T2 Image
TR	Short (500 msec)	Long (2000 msec)	Long (2000 msec)
TE	Short (15-20 msec)	Short (15-20 msec)	Long (120 msec)

INVERSION RECOVERY

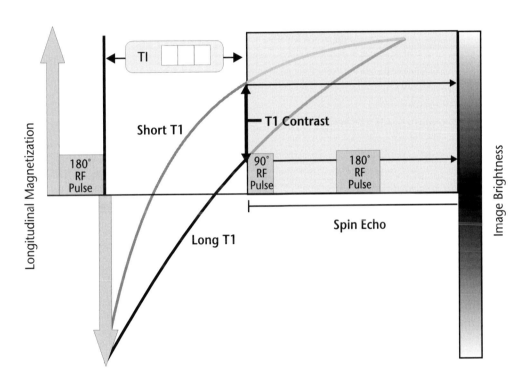

Figure 6-5. The inversion recovery method with TI set to produce an image with high T1 contrast.

why it is classified as one of the spin echo, rather than gradient echo, methods. An additional time interval is associated with the inversion recovery pulse sequence. The time between the initial 180° pulse and the 90° pulse is designated the Time after Inversion (TI). It can be varied by the operator and used as a contrast control.

T1 Contrast

The principal characteristic of many inversion recovery images is high T1 contrast. This occurs because the total longitudinal relaxation time is increased because it starts from the inverted state. There is more time for the T1 contrast to

T1 IMAGES

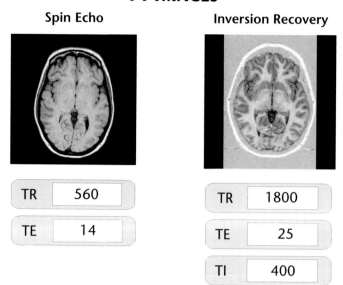

Spin Echo		Inversion Recovery	
TR	560	TR	1800
TE	14	TE	25
		TI	400

Figure 6-6. Comparison of T1 images produced by spin echo and inversion recovery methods.

develop. A T1 image produced by the inversion recovery method is compared to one produced by the spin echo method in Figure 6-6. Notice the significant difference in contrast. The use of the inversion method for other applications will be discussed in Chapter 8.

Mind Map Summary
Spin Echo Imaging Methods

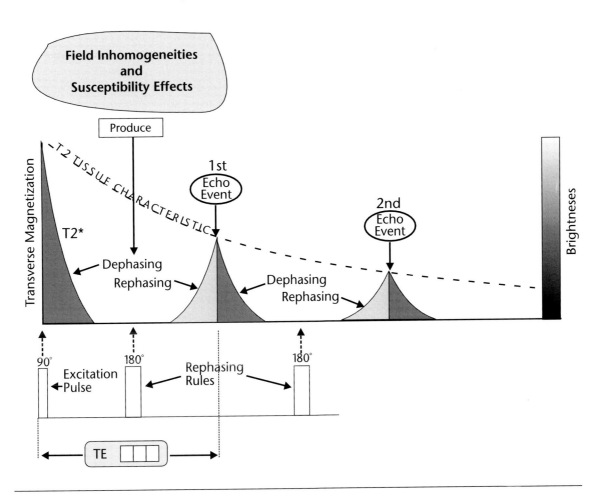

Spin echo is a technique used to produce an echo event by applying a 180° RF pulse to the dephased transverse magnetization. This compensates for the dephasing produced by field inhomogeneities and makes it possible to produce images that show the T2 characteristics of tissue. The time to the echo event, TE, is a protocol factor that can be adjusted to produce different weightings to the T2 contrast. When a short TE value is selected, the T2 effect is reduced, and the resulting image will be either a PD or T1-weighted image, depending on the selected TR value.

It is possible to use a series of 180° RF pulses within one cycle to produce multiple echo events, each with a different TE value. Both PD and T2-weighted images can be acquired in the same acquisition.

There is actually a family of spin echo methods that all use the spin echo process to create the echo event. These include the spin echo and inversion recovery methods as well as the GRASE method that uses both spin echoes and gradient echoes in the same acquisition.

7

Gradient Echo Imaging Methods

Introduction And Overview

It is possible to produce an echo event by applying a magnetic field gradient without a 180° RF pulse to the tissue as in the spin echo methods. There are several imaging methods that use the gradient echo technique to produce the RF signals and these make up the gradient echo family of methods.

The primary advantage of the gradient echo methods over the spin echo methods is that gradient echo methods perform faster image acquisitions. Gradient echo methods are generally considered to be among the faster imaging methods. They are also used in some of the angiographic applications because gradient echo generally produces bright blood, as

we will see in Chapter 12, as well as for functional imaging, as described in Chapter 13. One limitation of the gradient echo methods is they do not produce good T2-weighted images, as will be described later in this chapter. However, by combining the gradient and spin echo methods, this limitation can be overcome.

At this time we will develop the concept of gradient echo and then consider the specific gradient echo imaging methods and their characteristics.

The Gradient Echo Process

Transverse magnetization is present only when a sufficient quantity of protons are spinning in-phase in the transverse plane. As we have seen, the decay (relaxation) of transverse magnetization is the result of proton dephasing. We also recall that an RF signal is being produced any time there is transverse magnetization and the intensity of the signal is proportional to the level of magnetization.

With the spin echo technique we use an RF pulse to rephase the protons after they have been dephased by inherent magnetic field inhomogeneities and susceptibility effects within the tissue voxel. With the gradient echo technique the protons are first dephased, on purpose, by turning on a gradient and then rephased by reversing the direction of the gradient, as shown in Figure 7-1. A gradient echo can only be created when transverse magnetization is present. This can be either during the

free induction decay (FID) period or during a spin echo event. In Figure 7-1 the gradient echo is being created during the FID. Let us now consider the process in more detail.

First, transverse magnetization is produced by the excitation pulse. It immediately begins to decay (the FID process) because of the magnetic field inhomogeneities within each individual voxel. The rate of decay is related to the value of T2*. A short time after the excitation pulse a gradient is applied, which produces a very rapid dephasing of the protons and reduction in the transverse magnetization. This occurs because a gradient is a forced inhomogeneity in the magnetic field. The next step is to reverse the direction of the applied gradient. Even though this is still an inhomogeneity in

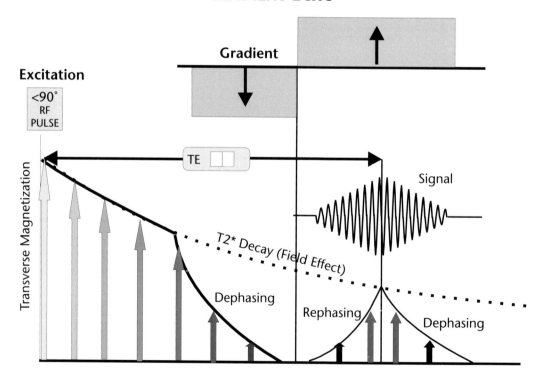

Figure 7-1. The gradient echo process using a magnetic field gradient to produce an echo event during the FID.

the magnetic field, it is in the opposite direction. This then causes the protons to rephase and produce an echo event. As the protons rephase, the transverse magnetization will reappear and rise to a value determined by the FID process. The gradient echo event is a rather well-defined peak in the transverse magnetization and this, in turn, produces a discrete RF signal.

The TE is determined by adjusting the time interval between the excitation pulse and the gradients that produce the echo event. TE values for gradient echo are typically much shorter than for spin echo, especially when the gradient echo is produced during the FID.

Small Angle Gradient Echo Methods

The gradient echo technique is generally used in combination with an RF excitation pulse that has a small flip angle of less than 90°. We will discover that the advantage of this is that it permits the use of shorter TR values and this, in turn, produces faster image acquisition.

One source of confusion is that each manufacturer of MRI equipment has given his gradient echo imaging methods different trade names. In this text we will use the generic name of *small angle gradient echo* (SAGE) method.

The SAGE method generally requires a shorter acquisition time than the spin echo methods. It is also a more complex method with respect to adjusting contrast sensitivity because the flip angle of the excitation pulse becomes one of the adjustable protocol imaging factors.

Excitation/Saturation-Pulse Flip Angle

We recall that the purpose of the excitation/ saturation pulse applied at the beginning of an imaging cycle is to convert or flip longitudinal magnetization into transverse magnetization. When a 90° pulse is used, all of the existing longitudinal magnetization is converted into transverse magnetization, as we have seen with the spin echo methods. The 90° pulse reduces the longitudinal magnetization to zero (i.e., complete saturation) at the beginning of each imaging cycle. This then means that a relatively long TR interval must be used to allow the longitudinal magnetization to recover to a useful value. The time required for the longitudinal magnetization to relax or to recover is one of the major factors in determining acquisition time. The effect of reducing TR when 90° pulses are used is shown in Figure 7-2. As the TR value is decreased, the longitudinal magnetization grows to a lower value and the amount of transverse magnetization and RF signal intensity produced by each pulse is decreased. The reduced signal intensity results in an increase in image noise as described in Chapter 10. Also, the use of short TR intervals with a 90° pulse (as in spin echo) cannot produce good PD or T2-weighted images.

One approach to reducing TR and increasing acquisition speed without incurring the disadvantages that have just been described is to use a pulse that has a flip angle of less than 90°. A small flip-angle (<90°) pulse converts only a

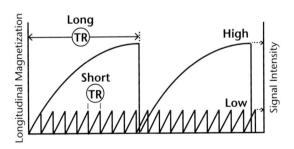

Figure 7-2. The effect of reducing TR on the recovery of longitudinal magnetization within a cycle and the resulting signal intensity when using 90° pulses.

fraction of the longitudinal magnetization into transverse magnetization. This means that the longitudinal magnetization is not completely destroyed or reduced to zero (saturated) by the pulse, as shown in Figure 7-3.

Reducing the flip angle has two effects that must be considered together. The effect that we have just observed is that the longitudinal magnetization is not completely destroyed and remains at a relatively high level from cycle to cycle, even for short TR intervals. This will increase RF signal intensity compared to the use of 90° pulses. However, as the flip angle is reduced, a smaller fraction of the longitudinal magnetization is converted into transverse magnetization. This has the effect of reducing signal intensity. The result is a combination of these two effects. This is illustrated in Figure 7-4. Here we see that as the flip angle is increased over the range from 0–90°, the *level* of longitudinal magnetization at the beginning of a cycle decreases. On the other hand, as the angle is increased, the *fraction* of this longitudinal magnetization that is converted into transverse magnetization increases and RF signal intensity increases. The combination of these two

LONGITUDINAL MAGNETIZATION

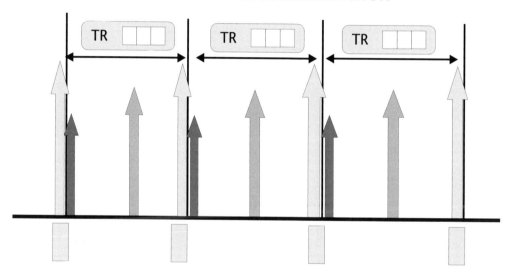

Small Flip Angle Pulses

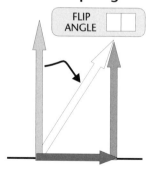

Figure 7-3. The effect of using small flip-angle pulses on longitudinal magnetization.

effects is shown in Figure 7-5. Here we see how changing flip angle affects signal intensity. The exact shape of this curve depends on the specific T1 value of the tissue and the TR interval. For each T1/TR combination there is a different curve and a specific flip angle that produces maximum signal intensity.

Let us now use Figure 7-6 to compare the magnetization of two tissues with different T1 values as we change flip angle. Contrast between the two tissues is represented by the difference in magnetization levels. At this point we are assuming a short TE and considering the contrast associated with only the longitudinal magnetization. The flip-angle range is divided into several specific segments as shown.

Contrast Sensitivity

With the SAGE method the contrast sensitivity for a specific tissue characteristic is controlled by three protocol factors. As with spin echo, TR and TE have an effect. However, the flip angle becomes the factor with the greatest effect on contrast. We will now see how changing flip angle can be used to select specific types of contrast with a basic gradient echo method.

T1 Contrast

Relatively large flip angles (45°–90°) produce T1 contrast. This is what we would expect because large flip angles (close to 90°) and short TR and TE values are similar to the factors used to

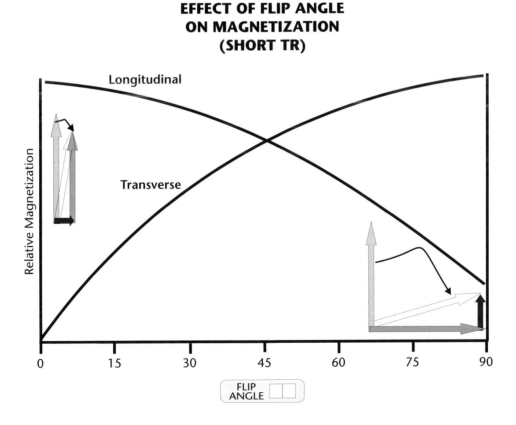

Figure 7-4. The effect of pulse flip angle on the level of both longitudinal and transverse magnetization after the pulse is applied.

EFFECT OF FLIP ANGLE
ON SIGNAL INTENSITY

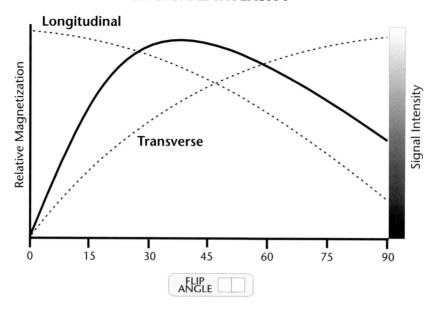

Figure 7-5. The relationship of signal intensity to flip angle.

EFFECT OF FLIP ANGLE
ON CONTRAST

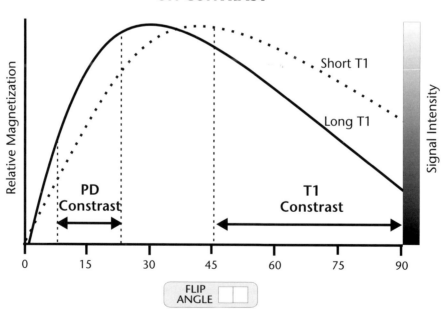

Figure 7-6. The effect of flip angle on contrast.

produce T1 contrast with the spin echo method. Here, with gradient echo, we observe a loss of T1 contrast as the flip angle is decreased significantly from 90°.

Low Contrast

There is an intermediate range of flip-angle values that produces very little, if any, contrast. This is the region in which the PD and T1 contrast cancel each other for many tissues, such as gray and white matter.

Proton Density (PD) Contrast

Relatively low flip-angle values produce PD contrast. As the flip angle is reduced within this region, there is a significant decrease in magnetization and the resulting signal intensity.

Up to this point we have observed generally how changing the flip angle of the excitation pulse affects signal intensity and contrast. In the SAGE imaging method the flip angle is one of the imaging factors that must be adjusted by the user. However, it becomes somewhat complex because the specific effect of flip angle is modified by the other imaging factors and techniques used to enhance a specific type of contrast.

T2 and T2* Contrast

We recall that T2 contrast is produced by the decay characteristics of transverse magnetization and that there are two different decay rates, T2 and T2*. The slower decay rate is determined by the T2 characteristics of the tissue. The faster decay is produced by small inhomogeneities within the magnetic field often related to variations in tissue susceptibility differences. This decay rate is determined by the T2* of the tissue environment. When a spin echo technique is used, the spinning protons are rephased, and the T2* effect is essentially eliminated. However, when a spin echo technique is not used, the transverse magnetization depends on the T2* characteristics. The gradient echo technique does not compensate for the inhomogeneity and susceptibility effect dephasing as the spin echo technique does. Also, without using a spin echo process the long TE values necessary to produce T2 contrast cannot be achieved. Therefore, a basic gradient echo imaging method is not capable of producing true T2 contrast. The contrast will be determined primarily by the T2* characteristics. The amount of T2* contrast in an image is determined by the selected TE value. In general, longer TE values (but short compared to those used in spin echo) produce more T2* contrast.

Contrast Enhancement

In addition to using combinations of TR, TE, and flip angle to control the contrast characteristics, some gradient echo methods have other features for enhancing certain types of contrast.

When SAGE methods are used with relatively short TR values, there is the possibility that some of the transverse magnetization created in one imaging cycle will carry over into the next cycle. This happens when the TR values are in the same general range as the T2 values of the tissue. SAGE methods differ in how they use the carry-over transverse magnetization.

A typical SAGE sequence is limited to one RF pulse per cycle. If additional pulses were used, as in the spin echo techniques, they would affect the longitudinal magnetization and upset its condition of equilibrium. However, because of the relatively short TR values it is possible for the repeating small-angle excitation pulses to produce a spin echo effect. This can occur only when the TR interval is not much longer than the T2 value of the tissue.

Associated with each excitation pulse, there are actually two components of the transverse magnetization. There is the FID produced by the immediate pulse and a spin echo component produced by the preceding pulses. The spin echo component is related to the T1 characteristics of the tissue. The FID component is related to the T2* characteristics. The contrast characteristics of the imaging method are determined by how these two components are combined. Different combinations are obtained by altering the location of the gradient echo event relative to the transverse magnetization and by turning the spin echo component on or off as described below.

Mixed Contrast

When both the FID and spin echo components are used, an image with mixed contrast characteristics will be obtained. This method produces a relatively high signal intensity compared to the methods described below.

Spoiling and T1 Contrast Enhancement

An image with increased Tl contrast is obtained by suppressing the spin echo component. This is known as *spoiling*. The spin echo component, which is a carryover of transverse magnetization from previous cycles, can be destroyed or spoiled by either altering the phase relationship of the RF pulses or by applying gradient pulses to dephase the spinning protons.

The basic SAGE method discussed up to this point permits faster (than spin echo) image acquisition because the TR can be set to shorter values. However, the gradient echo process can be used in methods that provide fast acquisition based on an entirely different principle. We will now consider methods that achieve their speed by filling many rows of k space during one acquisition cycle.

In Chapter 5 we saw that in the acquisition phase the signal data is being directed into k

space from which the image will be reconstructed. The k space is filled one row at a time. The number of rows in the k space for a specific image depends on the required image detail. The process that directs the signals into a specific row of k space is the spatial encoding function performed by one of the gradients. This will be described in Chapter 9. In conventional spin echo and SAGE imaging only one row of k space is filled with each imaging cycle. This is because there is only one echo signal produced per cycle that can be encoded to go to a specific row of k space. This means that the size of k space determines the minimum number of cycles that an acquisition must have. We are about to see some gradient echo methods that can fill many rows of k space in one imaging cycle. This is achieved by using the gradient echo process to produce many echo events from the transverse magnetization that is present during one cycle.

Echo Planar Imaging (EPI) Method

Echo planar is the fast gradient echo imaging method that is capable of acquiring a complete image in a very short time. However, it requires an MRI system equipped with strong gradients that can be turned on and off very rapidly. All systems do not have this capability. The EPI method consists of rapid, multiple gradient echo acquisitions executed during a single spin echo event. The unique characteristic of this method is that each gradient echo signal receives a different spatial encoding and is directed into a different row of k space. The actual spatial encoding process will be described in Chapter 9. Here we are considering only the general concept of EPI and how it achieves rapid acquisition.

The basic EPI method is illustrated in Figure 7-7. Here we see the actions that occur within one imaging cycle. The cycle usually begins with a spin echo pulse sequence that produces a spin echo event consisting of a period of transverse magnetization as described in Chapter 6. In conventional spin echo imaging we obtain only one signal (and fill one row of k space) from this period of magnetization. What we are about to do with EPI is to chop this one spin echo event into many shorter gradient echo events. The signals from each gradient echo event will receive different spatial encodings and fill different rows of k space.

It is possible to fill all of the k space and acquire a complete image in one cycle. This is described as *single shot* EPI. This is not always practical because it might place some limitations on the image quality that can be achieved and is also very demanding on the gradients. A more practical approach is to divide the acquisition into multiple shots (cycles) with each filling some fraction of the total k space.

The important factor is the number of gradient echo events created in each cycle. This is an adjustable protocol factor and is generally known as the EPI *speed factor*. This is the factor by which the acquisition time is reduced compared to a conventional method using the same TR value.

Gradient And Spin Echo (GRASE) Method

The GRASE method is, as the name indicates, a combination of gradient and spin echo methods. It provides the fast acquisition capability of gradient echo (EPI) with the superior

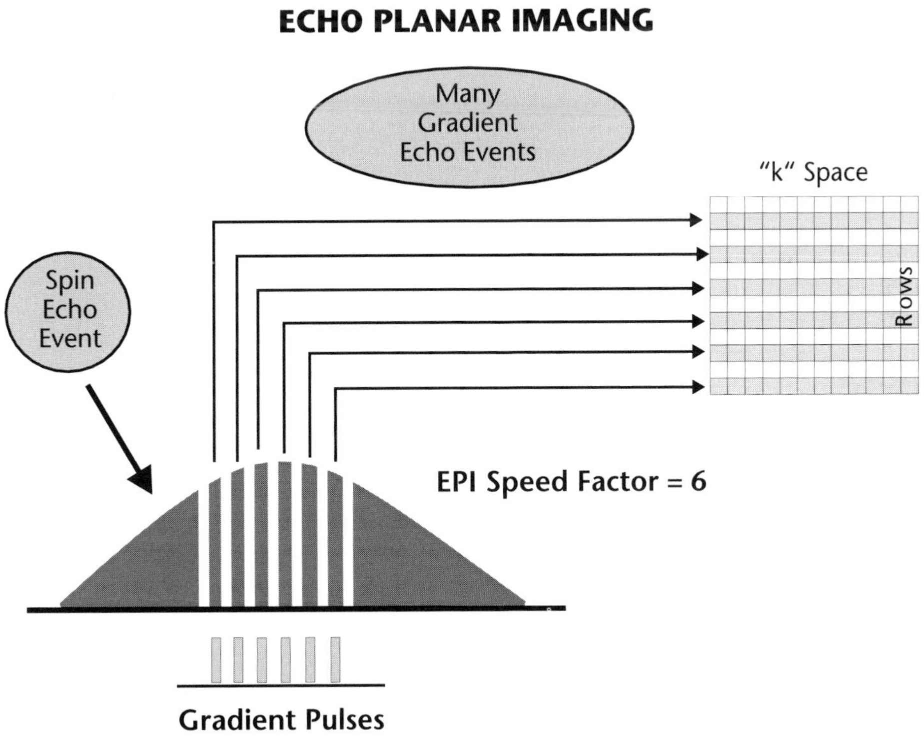

ECHO PLANAR IMAGING

Figure 7-7. The production of many gradient echo events within one imaging cycle with the echo planar imaging (EPI) method.

contrast characteristics of spin echo, including the ability to produce T2 images.

The GRASE method is illustrated in Figure 7-8 where we see the actions occurring within one imaging cycle. The basic cycle is a multiple spin echo as described in Chapter 5. The difference is that in conventional multiple echo, each of the echo events have different TE values and are used to form several images; typically, a PD and a T2 image with the same acquisition. Here the multiple spin echo is used for a different purpose. The multiple spin echoes are used to cover more of k space. As we see, each of the spin echo events is cut into many gradient echoes by the EPI process. This reduces the acquisition time by two factors: the total speed

factor is the number of multiple spin echoes multiplied by the EPI speed factor.

Magnetization Preparation

Both SAGE with short TR values and EPI can produce very rapid acquisitions. However, the short time intervals between the gradient echo events do not provide sufficient time for good longitudinal magnetization contrast (T1 or PD) to be formed. This problem is solved by "preparing" the magnetization and forming the contrast just one time at the beginning of the acquisition cycle, as shown in Figure 7-9. Two options are shown.

The longitudinal magnetization is prepared by applying either a saturation pulse, as in the

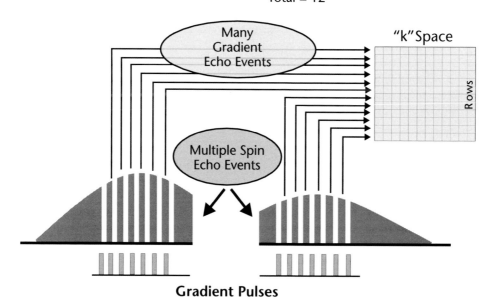

GRADIENT AND SPIN ECHO

Spin Echo Speed Factor = 2
EPI Speed Factor = 6
Total = 12

Figure 7-8. The use of the GRASE method to fill many rows of k space and produce a fast acquisition.

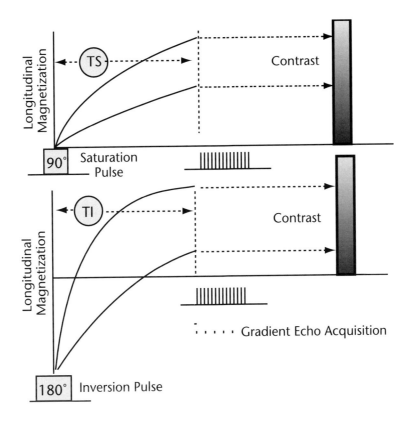

Figure 7-9. Using preparation pulses to produce longitudinal magnetization contrast prior to a rapid gradient echo acquisition.

spin echo method, or an inversion pulse, as in the inversion-recovery method. As the longitudinal magnetization relaxes, contrast is formed between tissues with different T1 and PD values. After a time interval [TI or TS (Time after Saturation)] selected by the operator, a rapid gradient echo acquisition begins.

The total acquisition time for this method is the time required by the acquisition cycles plus the TI or TS time interval.

Mind Map Summary
Gradient Echo Imaging Methods

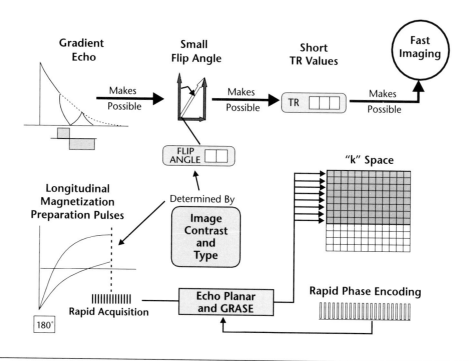

The common characteristic of the gradient echo imaging methods is that a magnetic field gradient is used to produce the echo event rather than a 180° RF pulse, as is used in the spin echo methods. One of the principal advantages of the gradient echo process is that it is a relatively fast imaging method.

By using a gradient, and not an RF pulse, to produce the echo event, it is possible to use saturation/excitation pulses with flip angles less than 90°; thereby all the longitudinal magnetization is not destroyed (saturated) at the beginning of each cycle. Because some longitudinal magnetization carries over from cycle to cycle, it is possible to reduce the TR value and still produce useful signal levels. The reduced TR values result in faster imaging. The flip angle of the RF pulse is an adjustable protocol factor that controls the type of contrast produced.

Echo planar imaging is a gradient echo method in which many echo events, each with a different phase encoding step, are created during each imaging cycle. This makes it possible to fill multiple rows of k space, which results in very fast imaging. GRASE is an imaging method that combines the principles of echo planar and fast (turbo) spin echo to produce rapid imaging acquisitions.

When very fast gradient echo methods are used, there is not sufficient time between the echo events for significant tissue relaxation and contrast to develop. Therefore, the desired contrast is developed at the beginning of the acquisition by applying either inversion or saturation "magnetization preparation" pulses. Then, when the desired contrast has developed, a rapid acquisition is performed.

8

Selective Signal Suppression

Introduction And Overview

There are many times when it is desirable to selectively suppress the signals from specific tissues or anatomical regions. This is done for a variety of reasons including the enhancement of contrast between certain tissues and the reduction of artifacts. During the acquisition process signals can be suppressed based on several properties of a tissue or fluid that make it different from other surrounding tissues. These include differences in T1 values, resonant frequencies, and molecular binding properties. Also, signals from specific anatomical regions can be suppressed or "turned off," usually to prevent interference with imaging in other areas. We will now see how these techniques are used.

Fat and fluid are two materials in the body that can produce very intense signals and brightness in images. This occurs with fat in T1 images and with fluid in T2 images. A possible problem is that these bright regions can reduce the visibility of other tissues and pathologic conditions in the area.

T1-Based Fat And Fluid Suppression

Let us recall that fat has very short T1 values (260 msec) and fluids have very long T1 values (2000 msec). These values are outside of the range of the T1 values of other tissues in the body and are separate and not mixed in with the others. This makes it possible to use T1 as a characteristic for the selective suppression of both fat and fluid.

STIR Fat Suppression

STIR is an inversion recovery method with the TI adjusted to selectively suppress the signals from fat. This uses the fact that fat has a relatively short T1 value and recovers its longitudinal magnetization faster than the other tissues after the inversion pulse. The important point here is that the magnetization of fat passes through the zero level before the other tissues, as shown in Figure 8-1. The TI interval is selected so that the "picture is snapped" by applying the excitation pulse at that time. Because the fat has no magnetization at that time, it will not produce a signal. Since this is achieved with relatively short values for TI, this method of fat suppression is often referred to as Short Time Inversion Recovery (STIR).

STIR is just the inversion recovery (IR) method with the TI set to a relatively low value. The description of the basic IR method in Chapter 6 shows how the factor TI is used to select the time at which the longitudinal magnetization "picture is snapped" and the magnetization is converted into image contrast. The ability to use this method to suppress the signals from fat is based on the fact that the longitudinal magnetization of fat passes through zero at a time before and separated from the other tissues. Setting the TI to measure the longitudinal magnetization at the time when fat is at zero produces no signal and fat will be dark in the image.

The best TI value to suppress the signals from fat depends on the T1 value of fat, which

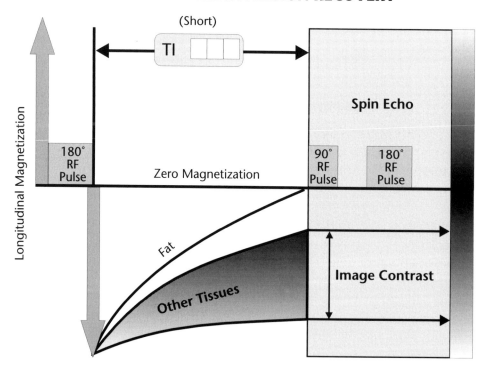

SHORT TIME INVERSION RECOVERY

Figure 8-1. The use of STIR to suppress signals from fat by setting TI to a value (short) that will image the longitudinal magnetization at the time when fat is relaxing through the zero level.

depends on the strength of the magnetic field. It will generally be in the range of 120 to 150 msec for field strengths in the 0.5 T to 1.5 T range.

Another consideration with STIR is that the TR must be set relatively long (1500–2000 msec), compared to a T1 image acquisition with spin echo using a TR value of approximately 500 msec. This additional time is required for the longitudinal magnetization to more fully recover after the excitation pulse and before the next cycle can begin.

Fluid Suppression

The suppression of signals from fluids can be achieved by using the IR Method with the TI set to relatively long values as shown in Figure 8-2. This works because the long T1 values of fluids are well separated from the T1 values of other tissues. By setting the TI to a long value as shown, the longitudinal magnetization is converted to transverse and the "picture is snapped" when the fluid is at a zero value. Fluids appear as dark regions in the image. When fluid suppression is used with a T2 image acquisition (long TE), the usually bright fluid is suppressed but other tissues with long T2 values, such as pathologic tissue, remain bright.

Acquisition time is a special concern with this method. That is because when long TI values are used, the TR values must also be long (5000–6000 msec) and that increases the acquisition time. For this reason, the practical

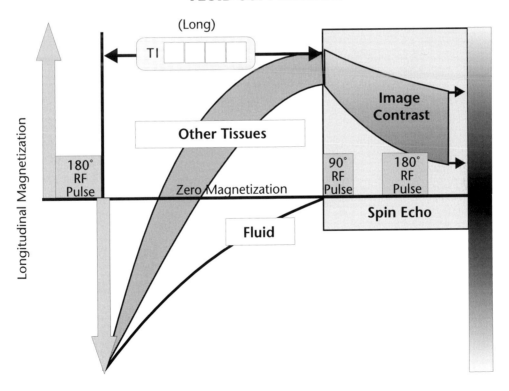

Figure 8-2. The suppression of fluid by selecting a long TI that will image the longitudinal magnetization at the time when fluid is relaxing through zero.

thing is to use this method with one of the fast acquisition techniques.

SPIR Fat Suppression

Spectral Presaturation with Inversion Recovery (SPIR) is a fat suppression technique which makes use of the fact that fat and the water content of tissues resonate at different frequencies (on the RF frequency spectrum) as described in Chapter 3. We must be careful not to confuse the two fat suppression methods, STIR and SPIR. As we have just seen, STIR uses the difference in *T1 values* to selectively suppress the signals from fat. Now with SPIR, we will use the differences in *resonant frequency* to suppress the fat signals. This technique is illustrated in Figure 8-3. The

unique feature of this method is that the imaging cycle begins with an inversion pulse that is applied at the fat resonant frequency. This selectively inverts the longitudinal magnetization of the fat without affecting the other tissues. The TI is set so that the spin echo excitation pulse is applied at the time when the fat longitudinal magnetization is passing through zero. This results in T1 and T2 images with the signals from fat removed.

The advantage of the SPIR method is that the contrast of tissues with relatively short T1 values is not diminished as it might be with the STIR method. For example, the use of gadolinium contrast media reduces the T1 value of the water component of tissue. These

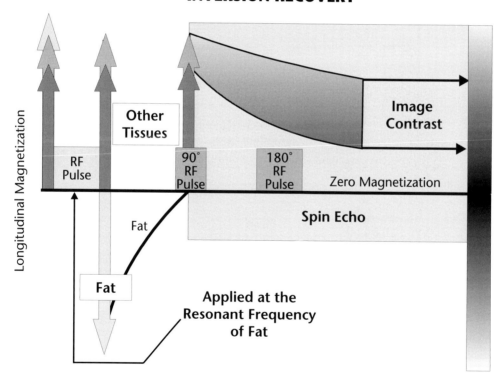

SPECTRAL PRESATURATION INVERSION RECOVERY

Figure 8-3. Suppressing the signals from fat by applying an inversion pulse tuned to the resonant frequency of fat so that it does not affect the other tissues.

short T1 value signals would be suppressed by STIR, but not by SPIR.

There are some precautions that must be observed when using SPIR. They relate to having very good magnetic field homogeneity. Recall that the resonant frequency is controlled by the field strength in each location. Therefore, for the RF suppression pulse to accurately suppress the fat magnetization over the image area, the fat must be resonating at precisely the same frequency. This requires a very homogeneous (within just a few parts per million) magnetic field. This is achieved by shimming the field before the acquisition, removing metal objects that might distort the field, and by using a relative small field of view.

An alternative to the SPIR method is to apply a saturation rather than an inversion pulse tuned to the fat resonant frequency. This is sometimes referred to as chemical saturation.

Magnetization Transfer Contrast (MTC)

Magnetization Transfer Contrast (MTC) is a technique that enhances image contrast by selectively suppressing the signals from specific tissues. The amount of suppression depends on a specific tissue's *magnetization transfer* characteristics. Maximum suppression is obtained for tissues that have a high level of magnetization transfer.

The MTC technique is illustrated in Figure 8-4. It is based on the principle that the protons in tissue are in different states of mobility, which we will designate as the "free" pool and the "bound" pool.

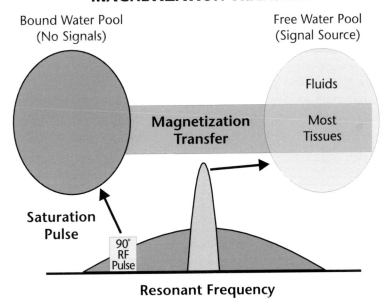

Figure 8-4. The use of magnetization transfer between different types of tissue to suppress selective signals.

Free Proton Pool

The protons that produce signals and are visible in MRI are not rigidly bound and might be considered to be "free" and in a general "semi-solid" structure. This environment produces relatively long T2 values (in comparison to the bound state) and a relatively narrow resonant frequency.

Bound Proton Pool

Most tissues also contain protons that are more rigidly bound and associated with more "solid" structures such as large macromolecules and membranes. These structures have very short T2 values. This means that the transverse magnetization decays before it can be imaged with the usual methods. Therefore, these protons do not contribute to the image. An important characteristic of these protons is that they have a much broader resonant frequency spectrum than the "free" protons.

Magnetization Transfer

Magnetization transfer is a process in which the longitudinal magnetization of one pool influences the longitudinal magnetization in the other pool. In other words, the longitudinal magnetizations of the two pools are coupled together but not to the same degree in all tissues. The MTC process makes use of this difference in coupling to selectively suppress the signals from certain tissues. This is how it is done.

Selective Saturation

The objective of this technique is to saturate and suppress selective signals from specific tissues to increase the contrast.

Prior to the beginning of the imaging acquisition cycle a saturation pulse is applied at a frequency that is different from the resonant frequency of the "free" protons. Therefore, it does not have a direct effect on the protons that are producing the signals. However, the saturation pulse is within the broader resonant frequency of the "bound" protons. It produces saturation of the longitudinal magnetization in the "bound" pool.

The effect of the saturation is now transferred to the longitudinal magnetization of the "free" pool by the magnetization transfer process. The key is that the transfer is not the same for all tissues. Only the tissues with a relatively high magnetization transfer coupling and a significant bound pool concentration will experience the saturation and have their signals reduced in intensity.

Fluids, fat, and bone marrow have very little, if any, magnetization transfer. Therefore, they will not experience the transferred saturation, and will remain relatively bright in the images.

Most other tissues have some, but varying degrees of, magnetization transfer. When the MTC technique is used, the saturation produced by the RF pulse applied to the "bound" protons will be transferred to the "free" protons, but only in those tissues that have a significant magnetization transfer capability. The result is that these tissues will be saturated to some degree and their signal intensities will be reduced.

Therefore, MTC is a way of enhancing contrast in an image by suppressing the signals from tissues that have a relatively high magnetization transfer. One example is to use MTC to reduce the brightness (signal intensity) of brain tissue so that the vascular structures will be brighter in angiography.

Regional Saturation

There are procedures in which it is desirable to suppress signals from specific anatomical regions. The two major applications of this are to reduce motion-induced artifacts, as described in Chapter 14, and to suppress the signals from blood that is flowing in a specific direction, as

REGIONAL SATURATION

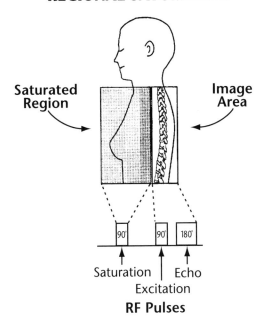

Figure 8-5. The application of a saturation pulse can be directed to a specific anatomical region to suppress undesirable signals from moving tissues.

discussed in Chapter 12. At this time we will consider the general technique, which is illustrated in Figure 8-5.

Let us recall that gradients are used to vary the magnetic field strength across a patient's body. In the presence of a gradient one region of the body is in a different field strength from another and is therefore tuned to a different resonant frequency. This makes it possible to apply RF pulses selectively to specific regions without affecting adjacent regions.

In Chapter 14 we will see that a major source of artifacts in MRI is the motion or movement of tissues and fluids. The motion produces errors in the spatial encoding of the signals that causes them to be displayed in the wrong location in the image. Signals from moving tissues and fluids are displayed as streaks, which are undesirable artifacts.

With the regional saturation technique the objective is to suppress selective signals originating from one region, usually the moving tissue or fluid, without affecting these signals in the region that is being imaged. The specific applications of this will be described in Chapter 14.

Prior to the imaging cycle pulse sequence, a saturation pulse is selectively applied to the region that is to be suppressed. The saturation pulse is given a frequency that is different from the frequency of the other imaging pulses. This is so that it will be tuned to the resonant frequency of the region that is to be suppressed. This region will have a resonant frequency different from the imaged area because of the presence of the gradient as described above.

The region that is saturated is a three-dimensional (3-D) volume or slab of tissue. It is important that the slab be properly positioned in relationship to the imaged area for best results.

The application of regional saturation to suppress artifacts will be discussed in more detail when we consider artifacts in Chapter 14.

Mind Map Summary
Selective Signal Suppression

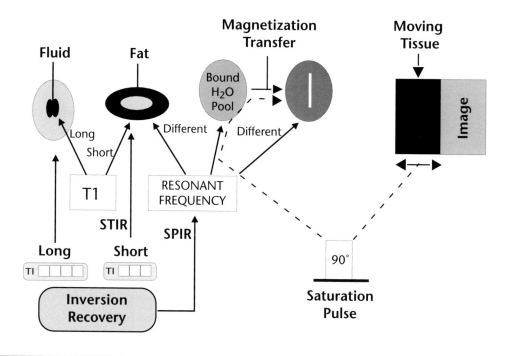

It is often desirable to suppress the signals and resulting brightness of selected tissues or anatomical regions to improve visibility of other tissues or general image quality. It is possible to selectively suppress signals from specific tissues if the tissues are significantly different from the other tissues in terms of some MR characteristic.

Signals from fat, generally very bright in T1 images, can be suppressed with two techniques. Because fat has a very short T1 value compared to other tissues, it can be suppressed with the STIR method, an inversion recovery method in which the TI is set to snap the picture when the magnetization of fat is passing through the zero level. The resonant frequency of fat molecules is slightly different from water molecules because of the chemical shift effect. The SPIR method makes use of this by applying an RF pulse at the fat frequency to reduce the fat magnetization to the zero level at the beginning of each imaging cycle.

Signals from fluid can be suppressed by using an inversion recovery method with the TI set to a long value. This works because fluids have long T1 values and the fluid's magnetization passes through the zero level significantly later and separate from that of tissues. The MTC technique can be used to reduce signal intensity from tissues that have a relatively high magnetization transfer characteristic. This can be used to enhance image contrast.

Saturation pulses can be selectively applied to specific anatomical regions to suppress any signals that could occur from tissues or fluids in that region. This is useful for reducing motion artifacts and also for reducing the signals from flowing blood in specific anatomical regions.

9

Spatial Characteristics of the Magnetic Resonance Image

Introduction And Overview

The MR image formation process subdivides a section of the patient's body into a set of slices and then each slice is cut into rows and columns to form a matrix of individual tissue voxels. This was introduced first in Chapter 1 and illustrated in Figure 1-3. The RF signal from each individual voxel must be separated from all of the other voxels and its intensity displayed in the corresponding image pixel, as shown in Figure 9-1. This is achieved by encoding or addressing the signals during the acquisition phase and then, in effect, delivering the signal intensities to the appropriate pixels which have addresses within the image during the reconstruction phase. Because there are two dimensions, or directions, in an image, two different methods of encoding must be used. This is analogous to mail that must have both a street name and a house number in the address. We are about to see that the two methods of addressing the signals are called *frequency-encoding* and *phase-encoding*. One method is applied to one direction in the image and the other method is used to address in the other direction.

This two-step process consisting of the *signal acquisition* phase followed by the *image reconstruction* phase is illustrated in Figure 9-2. Different actions happen in these two phases that must be considered when setting up an imaging procedure.

SPATIAL CHARACTERISTICS

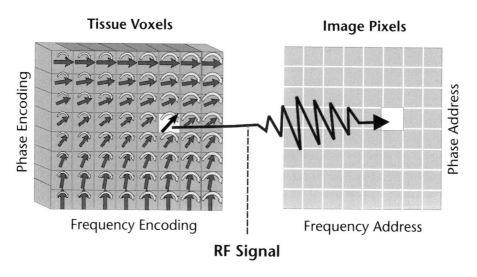

Figure 9-1. The relationship of tissue voxels to image pixels.

MRI IMAGE PRODUCTION

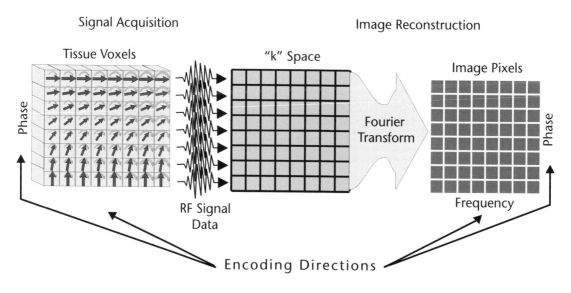

Figure 9-2. The two phases—signal acquisition and image reconstruction—that are required to produce an MR image.

Signal Acquisition

During the acquisition phase the RF signals are emitted by the tissue and received by the RF coils of the equipment. During this process the signals from the different slices and voxels are given distinctive frequency and phase characteristics so that they can be separated from the other signals during image reconstruction. The acquisition phase consists of an imaging cycle that is repeated many times. The time required for image acquisition is determined by the time TR, which is the duration of one cycle or its repetition time, and the number of cycle repetitions. The number of cycles is determined by the image quality requirements. In general, the quality of an image can be improved by increasing the number of acquisition cycles. This is considered in much more detail in Chapter 10.

The result of the image acquisition process is a large amount of data collected and stored in computer memory. At this point the data represent RF signal intensities characterized by the two characteristics, *frequency* and *phase*. The concept of frequency and phase will be developed later. At this point in the process the data are not yet in the form of an image but are located in k space. The data will later be transformed into image space by the reconstruction process.

Image Reconstruction

Image reconstruction is a mathematical process performed by the computer. It transforms the data collected during the acquisition phase into an image. We can think of reconstruction as the process of sorting the signals collected during the acquisition and then delivering them to the appropriate image pixels. The mathematical process used is known as *Fourier transformation.* Image reconstruction is typically much faster than image acquisition and requires very little, if any, control by the user.

Image Characteristics

The most significant spatial characteristic of an image is the size of the individual tissue voxels. Voxel size has a major effect on both the detail and noise characteristics of the image. The user can select the desired voxel size by adjusting a combination of imaging factors, as described in Chapter 10.

Gradients

The spatial characteristics of an MR image are produced by actions of the gradients applied during the acquisition phase. Magnetic field gradients are used first to select slices and then give the RF signals the frequency and phase characteristics that create the individual voxels.

As we will see later, a gradient in one direction is used to create the slices, and then gradients in the other directions are used to cut the slices into rows and columns to create the individual voxels. However, these functions can be interchanged or shared among the different gradient coils to permit imaging in any plane through the patient's body.

The functions performed by the various gradients usually occur in a specific sequence. During each individual image acquisition cycle the various gradients will be turned on and off at specific times. As we will see later, the gradients are synchronized with other events such as the application of the RF pulses and the acquisition of the RF signals.

Slice Selection

There are two distinct methods used to create the individual slices. The method of *selective excitation* actually creates the slice during the acquisition phase. An alternative method is to acquire signals from a large volume of tissue (like an organ) and then create the slices during the reconstruction process. These are often referred to as 2-D (volume) and 3-D (volume) acquisitions. However, each produces data that are reconstructed into slice images. Both methods have advantages and disadvantages, which will be described later.

Selective Excitation

The first gradient action in a cycle defines the location and thickness of the tissue slice to be imaged. We will illustrate the procedure for a conventional transaxial slice orientation. Other orientations, such as sagittal, coronal, and angled combinations, are created by interchanging and combining gradient directions.

Slice selection using the principle of selective excitation is illustrated in Figure 9-3. When a magnetic field gradient is oriented along the patient axis, each slice of tissue is in a different field strength and is tuned to a different resonant frequency. Remember, this is because the resonant frequency of protons is directly proportional to the strength of the magnetic field at the point where they are located. This slice selection gradient is present whenever RF pulses are applied to the body. Since RF pulses contain frequencies within a limited range (or bandwidth), they can excite tissue only in a specific slice. The location of the slice can be changed or moved along the gradient by using a slightly different RF pulse frequency. The thickness of the slice is determined by a combination of two factors: (1) the strength, or steepness, of the gradient, and (2) the range of frequencies, or bandwidth, in the RF pulse.

Multi-Slice Imaging

In most clinical applications, it is desirable to have a series of images (slices) covering a specific anatomical region. By using the multi-slice mode, an entire set of images can be acquired simultaneously. The basic principle is illustrated in Figure 9-4. The slices are separated by applying the RF pulses and detecting the signals from the different slices at different times, in sequence, during each imaging cycle.

When the slice selection gradient is turned on, each slice is tuned to a different resonant frequency. A specific slice can be selected for

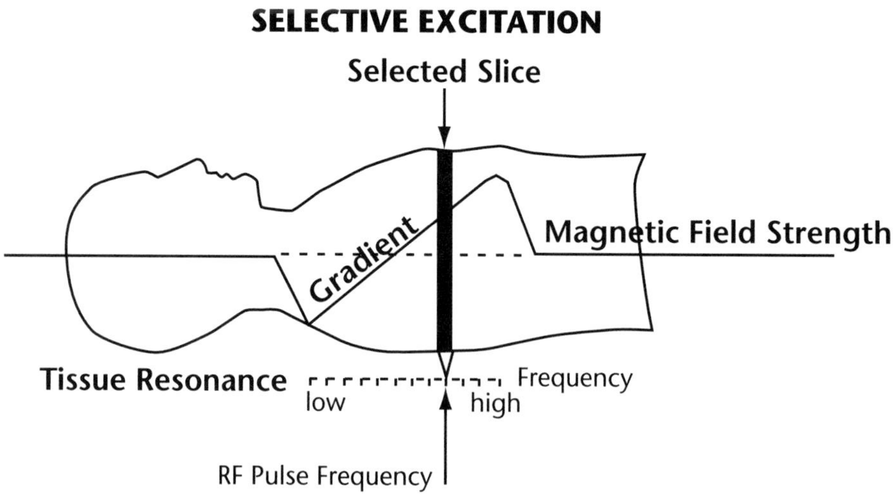

SELECTIVE EXCITATION

Selected Slice

Magnetic Field Strength

Gradient

Tissue Resonance   Frequency
low   high

RF Pulse Frequency

Figure 9-3. The use of a gradient to tune a specific slice so that it can be selectively excited by an RF pulse.

excitation by adjusting the RF pulse frequency to correspond to the resonant frequency of that slice. The process begins by applying an excitation pulse to one slice and collecting the echo signal. Then, while that slice undergoes longitudinal relaxation before the next cycle can begin, the excitation pulse frequency is shifted to excite another slice. This process is repeated to excite and collect signals from the entire set of slices at slightly different times within one TR interval.

The advantage of multi-slice imaging is that a set of slices can be imaged in the same time as a single slice. The principal factor that limits the number of slices is the value of TR. It takes a certain amount of time to excite and then collect the signals from each slice. The maximum number of slices is the TR value divided by the time required for each slice. This limitation is especially significant for T1-weighted images that use relatively short TR values.

A factor to consider when selecting the slicing mode is that multiple slice selective excitation cannot produce the contiguous slices that the volume acquisition technique can. With selective excitation there is the possibility that when an RF excitation/saturation pulse is applied to one slice of tissue, it will also produce some effect in an adjacent slice. This is a reason for leaving gaps between slices during the acquisition.

Volume Acquisition

Volume (3-D) image acquisition has the advantage of being able to produce thinner and more contiguous slices. This is because of the process used to slice the tissue. Rather than producing each slice during the acquisition phase, the slicing is done during the reconstruction phase using the process of phase-encoding. The actual process of phase-encoding will be described later in this chapter. At this time we only consider how it is used for slicing. With this method, no gradient is present when the RF pulse is applied to the tissue. Since all tissue within an anatomical region, such as the head, is tuned to the same resonant frequency, all tissues are

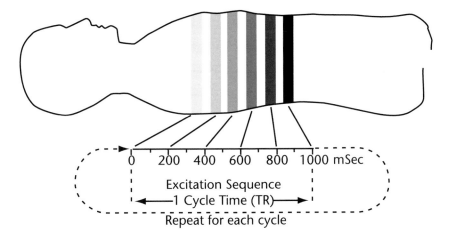

MULTIPLE SLICE IMAGING

Figure 9-4. Multiple slice imaging applies pulses to and produces signals from different slices within one imaging cycle.

3-D VOLUME ACQUISITION

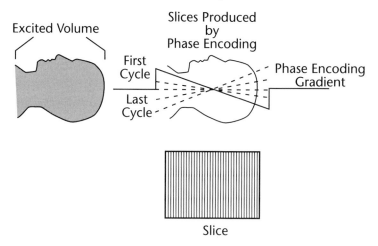

Figure 9-5. The 3-D volume acquisition process uses the phase-encoding process to produce thin slices.

excited simultaneously. The next step, as illustrated in Figure 9-5, is to apply a phase-encoding gradient in the slice selection direction. In volume imaging, phase-encoding is used to create the slices in addition to creating the voxel rows as described below. The phase-encoding gradient used to define the slices must be stepped through different values, corresponding to the number of slices to be created. At each gradient setting, a complete set of imaging cycles must be executed. Therefore, the total number of cycles required in one acquisition is multiplied by the number of slices to be produced. This has the disadvantage of causing 3-D volume acquisitions to have a relatively long acquisition time compared to 2-D multiple slice acquisitions. That is why this type of acquisition is often used with one of the faster imaging methods.

The primary advantage of volume imaging is that the phase-encoding process can generally produce thinner and more contiguous slices than the selective excitation process used in 2-D slice acquisition. The primary disadvantage is longer acquisition times.

Frequency Encoding

A fundamental characteristic of an RF signal is its frequency. Frequency is the number of cycles per second of the oscillating signal. The frequency unit of Hertz (Hz) corresponds to one cycle per second. Radio broadcast stations transmit signals on their assigned frequency. By tuning our radio receiver to a specific frequency we can select and separate from all other signals the specific broadcast we want to receive. In other words, the radio broadcasts from all of the stations in a city are frequency encoded. The same process (frequency-encoding) is used to cause voxels to produce signals that are different and can be used to create one dimension of the image.

Let us review the concept of RF signal production by voxels of tissue, as shown in Figure 9-6. RF signals are produced only when transverse magnetization is present. The unique characteristic of transverse magnetization that produces the signal is a spinning magnetic effect, as shown. The transverse magnetization spins around the axis of the magnetic

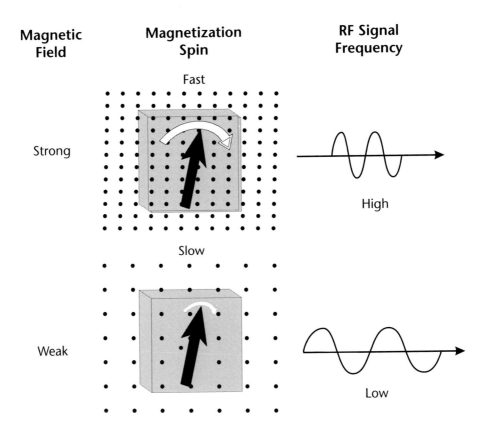

Figure 9-6. The effect of field strength on the frequency of RF signals produced by transverse magnetization.

field. A spinning magnet or magnetization in the vicinity of a coil forms a very simple electric generator. It generates one cycle for each revolution of the magnetization. When the magnetization is spinning at the rate of millions of revolutions per second, the result is an RF signal with a frequency in the range of Megahertz (MHz).

Resonant Frequency

The frequency of the RF signal is determined by the spinning rate of the transverse magnetization. This, in turn, is determined by two factors, as was described in Chapter 3. One factor is the specific magnetic nuclei (usually protons) and the other is the strength of the magnetic field

in which the voxel is located. When imaging protons, the strength of the magnetic field is the factor used to vary the resonant frequency and the corresponding frequency of the RF signals. In Figure 9-6 we see two voxels located in different strength fields. The result is that they produce different frequency signals.

Figure 9-7 shows the process of frequency encoding the signals for a row of voxels. In this example, a gradient is applied along the row. The magnetic field strength is increased from left to right. This means that each voxel is located in a different field strength and is resonating at a frequency different from all of the others. The resonant and RF signal frequencies increase from the left to right as shown.

FREQUENCY ENCODING

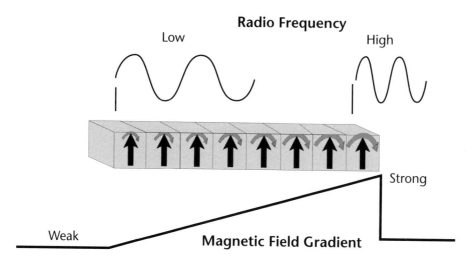

Figure 9-7. The frequency encoding of a row of voxels within a slice.

The frequency-encoding gradient is on at the time of the echo event when the signals are actually being produced. The signals from all of the voxels in a slice are produced simultaneously and are emitted from the body mixed together to form a composite signal at the time of the echo event. The individual signals will be separated later by the reconstruction process to form the voxels.

Phase-Encoding

Phase is a relationship between one signal and another, as illustrated in Figure 9-8. Here we see two voxels producing RF signals. The transverse magnetization is spinning at the same rate and producing signals that have the same frequency. However, we notice that one signal is more advanced in time or is out of step with the other. In other words, the two signals are out of phase. The significance of voxel-to-voxel phase in MRI is that it can be used to separate signals and create one dimension in the image.

A phase difference is created by temporarily changing the spinning rate of the magnetization

of one voxel with respect to another. This happens when the two voxels are located in magnetic fields of different strengths. This can be achieved by turning on a gradient, as shown in Figure 9-9.

Let us begin the process of phase encoding by considering the column of voxels shown in the illustration. We are assuming that all voxels have the same amount of transverse magnetization and that the magnetization is spinning in-phase at the time just prior to the phase-encoding process.

When the phase-encoding gradient is turned on, we have the condition illustrated with the center column of voxels. The strength of the magnetic field is increasing from bottom to top. Therefore, the magnetization in each voxel is spinning at a different rate with the speed increasing from bottom to top. This causes the magnetization from voxel to voxel to get out of step or produce a phase difference. The phase-encoding gradient remains on for a short period of time and then is turned off. This leaves the condition represented by the

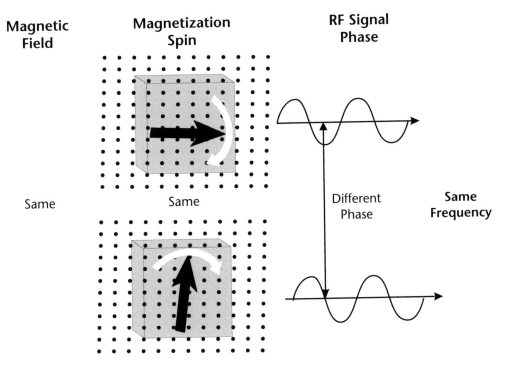

Figure 9-8. The concept of phase between the signals from two voxels.

PHASE ENCODING

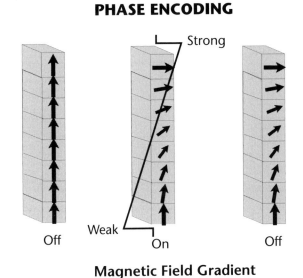

Magnetic Field Gradient

Figure 9-9. Phase-encoding produced by turning on a gradient for a short time, and then turning it off. The phase difference remains.

column of voxels on the right. This is the condition that exists at the time of the echo event when the signals are actually produced. As we see, the signals from the individual voxels are different in terms of their phase relationship. In other words, the signals are phase-encoded. All of the signals are emitted at the same time and mixed together as a composite echo signal. Later, the reconstruction process will sort the individual signal components.

Phase-encoding is the second function performed by a gradient during each cycle, as shown in Figure 9-10. During each pass through an imaging cycle, the phase-encoding gradient is stepped to a slightly different value.

The signals acquired with each phase-encoding *gradient strength* fills one row of k space. This is a very important point that should be emphasized: *Each row of k space is reserved for signals with a specific degree of phase-encoding.* The degree of phase-encoding is determined by the strength and duration of the phasing gradient applied during each cycle. Therefore, the phase-encoding process must be

repeated depending on the size of k space and that is determined by the image matrix size in the phase-encoded direction.

One MRI phase-encoding step produces a composite signal from all voxels within a slice. The difference from one step to another is that individual voxel signals have a different phase relationship within the composite signal.

To reconstruct an image by the conventional 2-D Fourier transformation method, one composite signal, or phase-encoded step, must be collected for each voxel to be created in the phase-encoding direction. Therefore, the minimum number of steps required to produce an image is determined by the size of the image matrix and k space. It takes 256 phase-encoding steps to produce an image with a 256×256 matrix.

The Gradient Cycle

We have seen that various gradients are turned on and off at specific times within each imaging cycle. The relationship of each gradient to the other events during an imaging cycle is shown in Figure 9-10. The three gradient activities are:

1. The slice selection gradient is on when RF pulses are applied to the tissue. This limits magnetic excitation, inversion, and echo formation to the tissue located within the specific slice.
2. The phase-encoding gradient is turned on for a short period in each cycle to produce a phase difference in one dimension of the image. The strength of this gradient is changed slightly from one cycle to another to fill the different rows of k space needed to form the image.
3. The frequency-encoding gradient is turned on during the echo event when the signals are actually emitted by the tissue. This causes the different voxels to emit signals with different frequencies.

GRADIENTS

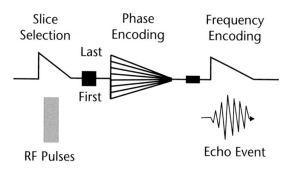

One Imaging Cycle

Figure 9-10. The relationship of the three gradient actions—slice selection, phase-encoding, and frequency-encoding— to each other and to the RF pulses and signals. They are applied in different directions.

MAGNETIC RESONANCE IMAGING

Because of the combined action of the three gradients, the individual voxels within each slice emit signals that are different in two respects—they have a phase difference in one direction and a frequency difference in the other. Although these signals are emitted at the same time, and picked up by the imaging system as one composite signal at the time of the echo event in each cycle, the reconstruction process can sort the signals into the respective components and display them in the correct image pixel locations.

Image Reconstruction

The next major step in the creation of an MR image is the reconstruction process. Reconstruction is the mathematical process performed by the computer that converts the collected

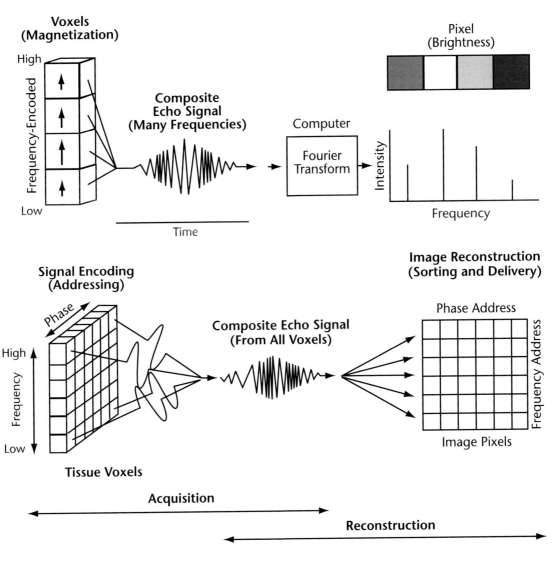

Figure 9-11. The concept of signal encoding (addressing) and image reconstruction (sorting and delivery).

signals in k space into an actual image. There are several reconstruction methods, but the one used for most clinical applications is the 2-D Fourier transformation.

It is a mathematical procedure that can sort a composite signal into individual frequency and phase components. Since each voxel in a row emits a different signal frequency and each voxel in a column a different phase, the Fourier transformation can determine the location of each signal component and direct it to the corresponding pixel.

Let us now use the concept illustrated in Figure 9-11 to summarize the spatial characteristics of the MR image. We will use a postal analogy for this purpose.

In the image each column of pixels has a phase address corresponding to different street names. Each row of pixels has a frequency address corresponding to house numbers. Therefore, each individual pixel has a unique address consisting of a combination of frequency and phase values analogous to a street name and house number.

The frequency- and phase-encoding process during acquisition "writes" an address on the signal from each voxel. These signals are mixed together and collected in a "post-office" called k space. The signals ("mail") are then sorted by the Fourier transform process and hopefully delivered to the correct pixel address in the image.

In Chapter 14 we will see that if a voxel of tissue moves during the acquisition process, it might not receive the correct phase address and the signal will be delivered to the wrong pixel. This creates ghost images and streak artifacts in the phase-encoded direction.

The chemical-shift artifact is caused by the difference in signal frequency between tissues containing water and fat. When it is present in an image, signals from the water components and fat will be offset by a few pixels. We will see how this is controlled in Chapter 14.

Mind Map Summary
Spatial Characteristics of the Magnetic Resonance Image

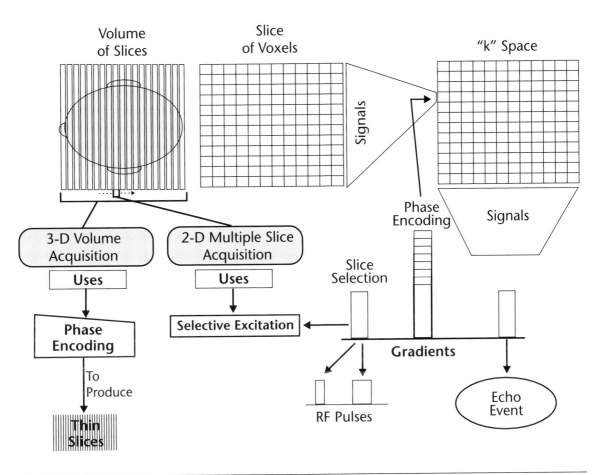

During an MRI procedure a section of a patient's body is first divided into a set of slices, and then each slice is divided into a matrix of voxels. These actions are produced by the gradients.

Two methods can be selected to produce the slices. The most common method, 2-D multiple slice acquisition, applies a gradient so that an individual slice is tuned to a resonant frequency different from the other slice positions. This gradient is turned on when the RF pulses are applied. Therefore, only the tissue in a specific slice is excited and goes through the process to produce signals. An alternate method, 3-D volume acquisition, uses phase-encoding to produce slices. It is generally capable of producing thinner, more contiguous slices.

Two different methods are used to cut a slice into voxels. Phase-encoding is used in one direction, and frequency-encoding in the other. Phase-encoding is produced by applying a gradient to the transverse magnetization during each imaging cycle. To produce sufficient phase-encoding information to permit image reconstruction, many different phase-encoding gradient strengths must be used. In the typical imaging procedure the phase-encoding gradient strength is changed from cycle to cycle. The strength of the phase-encoding gradient, in effect, directs the signal data into a specific row of k space. All the rows of k space must be filled with data before the image reconstruction can be performed. The number of rows of k space is one of the factors that determine how many imaging cycles must be used, which, in turn, affects image acquisition time.

Frequency-encoding is produced by applying a gradient at the time of the echo event during each cycle.

10

Image Detail and Noise

Introduction And Overview

Two characteristics of the MR image that reduce the visibility of anatomical structures and objects within the body are *blurring* and *visual noise*. These were introduced in Chapter 1 as image quality characteristics. Both image blurring and visual noise are undesirable characteristics that collectively reduce the overall quality of an image and the objects in the image as illustrated in Figures 1-6 and 1-7. In an image, the combined effects of blur and noise produce a "curtain of invisibility" that extends over some objects based on object characteristics. This is shown in Figure 10-1, where we see objects arranged according to two characteristics. In the horizontal direction,

the objects are arranged according to size. Decreasing object size corresponds to increasing detail. In the vertical direction, the objects are arranged according to their contrast. The object in the lower left is both large and has a high level of contrast. This is the object that would be most visible under a variety of imaging conditions. The object that is always the most difficult to see is the small, low contrast object, which in Figure 10-1 would be located in the upper right corner.

In every imaging procedure we can assume that some potential objects within the body will not be visible because of the blurring and noise in the image. This loss of visibility is represented by the "curtain" or area of invisibility indicated in Figure 10-1. The location of the boundary between the visible and invisible objects, often referred to as a *contrast-detail curve*, is determined by the amount of blurring and noise associated with a specific imaging procedure. In general, blurring reduces the visibility of anatomical detail or other small objects that are located in the lower right region. Visual noise reduces the visibility of low contrast objects located in the upper left region.

The imaging protocol determines the boundary of visibility by altering the amount of blurring and noise. These two characteristics are determined by the combination of many adjustable imaging factors. It is a complex process because the factors that affect visibility of detail (blurring) also affect noise, but in the opposite direction. As we will see when a protocol is changed to improve visibility of detail, the noise is increased. Another point to consider is that several of the factors that have an

effect on both image detail and noise also affect image acquisition time, which will be discussed in Chapter 11. Therefore, when formulating an imaging protocol one must consider the multiple effects of the imaging factors and then select factor values that provide an appropriate compromise and an optimized acquisition for a specific clinical study with respect to detail (blurring), noise, and acquisition speed.

We will now consider the many factors that have an effect on the characteristics of image detail and noise.

Image Detail

The ability of a magnetic resonance image to show detail is determined primarily by the size of the tissue voxels and corresponding image pixels. Pixel size can be changed without major tradeoffs. However, as we are about to observe, there are significant effects of changing voxel

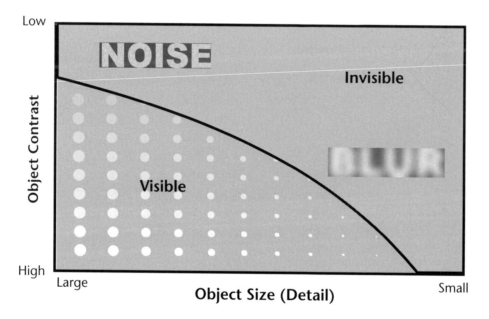

Figure 10-1. The impact of image noise and blurring on object visibility. Noise reduces visibility of low contrast objects. Blur reduces visibility of small objects.

size that must be considered. The real challenge is selecting a voxel size that is optimum for a specific clinical procedure.

In principle, all structures within an individual voxel are blurred together and represented by the signal intensity representing that voxel. It is not possible to see details within a voxel, just the voxel itself. When we view an MR image, we are actually looking at an image of a matrix, or array, of the voxels. We usually do not see the individual voxels because they are so small and they might be interpolated into even smaller image pixels. However, even if we do not see the individual voxels, their size determines the anatomical detail that we can see. The amount of image blurring is determined by the dimensions of the individual voxels.

Three basic imaging factors determine the dimensions of a tissue voxel, as illustrated in Figure 10-2. The dimension of a voxel in the plan of the image is determined by the ratio of the field of view (FOV) and the size of the matrix. Both of these factors can be used to adjust image detail. The thickness of the slice is a factor in voxel signal intensity.

The selection of the FOV is determined primarily by the size of the body part being imaged. One problem that can occur is the appearance of *foldover artifacts* when the FOV is smaller than the actual body section. However, there are artifact suppression techniques that can be used to reduce this foldover problem, as described in Chapter 14. The maximum useful FOV is usually limited by the dimensions and characteristics of the RF coil. The important thing to remember is that smaller image FOVs and smaller voxels produce better visibility of detail.

Matrix size refers to the number of voxels in the rows or columns of the matrix. The matrix size is a protocol factor selected by the

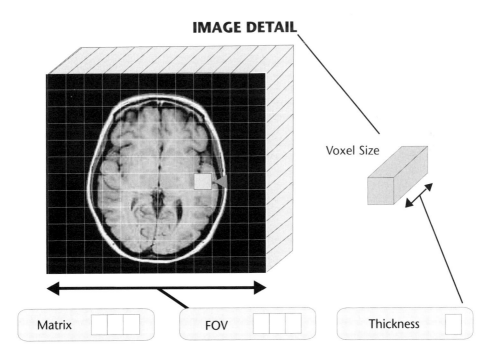

Figure 10-2. Voxel size and detail in MR images is determined by the values selected for the three protocol factors: FOV, matrix size, and slice thickness.

operator before the imaging procedure. Typical matrix dimensions are in the range of 128 to 512 mm.

Noise Sources

Random RF energy can be generated by thermal activity within electrical conductors and circuit components of the receiving system. In principle, the patient's body is a component of the RF receiving system. Because of its mass, it becomes the most significant source of image noise in most imaging procedures. The specific noise source is the tissue contained within the sensitive FOV of the RF receiver coils. Some noise might be generated within the receiver coils or other electronics, but it is usually much less than the noise from the patient's body.

Many devices in the environment produce RF noise or signals that can interfere with MRI. These include radio and TV transmitters, electrosurgery units, fluorescent lights, and computing equipment. All MR units are installed with an RF shield, as described in Chapter 2, to reduce the interference from these external sources. External interference is not usually a problem with a properly shielded unit. When it does occur, it generally appears as an image artifact rather than the conventional random noise pattern.

Signal-To-Noise Considerations

Image quality is not dependent on the absolute intensity of the noise energy but rather the amount of noise energy in relation to the image signal intensity. Image quality increases in proportion to the signal-to-noise ratio. When the intensity of the RF noise is low in proportion to the intensity of the image signal, the noise has a low visibility. In situations where the signals are relatively weak, the noise becomes much more visible. The principle is essentially the same as with conventional TV reception. When a strong signal is received, image noise (snow) is generally not visible; when one attempts to tune in to a weak TV signal from a distant station, the noise (noise) becomes significant.

In MRI, the loss of visibility resulting from the noise can be reduced by either *reducing the noise intensity* or *increasing the intensity of the signals*. This is illustrated in Figure 10-3. Let us now see how this can be achieved.

Voxel Size

One of the major factors that affects signal strength is the volume of the individual voxels. The signal intensity is proportional to the total number of protons contained within a voxel. Large voxels, that contain more protons, emit stronger signals and result in less image noise. Unfortunately, as we have just discovered, large voxels reduce image detail. Therefore, when the factors for an imaging procedure are being selected, this compromise between signal-to-noise ratio and image detail must be considered. The major reason for imaging relatively thick slices is to increase the voxel signal intensity and it also allows shorter TE values.

Field Strength

The strength of the RF signal from an individual voxel generally increases in proportion to the square of the magnetic field strength. However, the amount of noise picked up from the patient's body often increases with field strength because of adjustments in the bandwidth factor for the higher fields. This is described in Chapter 14. Because of differences in system design, no one precise relationship between signal-to-noise ratio and magnetic

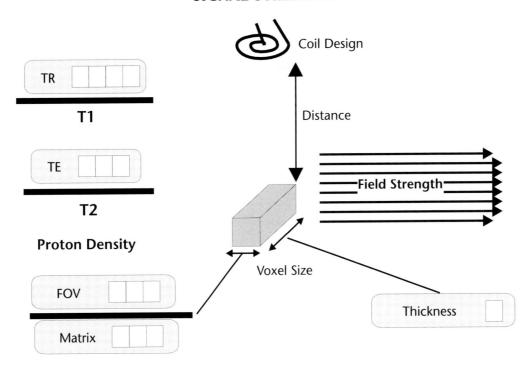

Figure 10-3. Factors that affect the signal-to-noise ratios in MR images.

field strength applies to all systems. In general, MRI systems operating at relatively high field strengths produce images with higher signal-to-noise ratios than images produced at lower field strengths, when all other factors are equal.

Tissue Characteristics

Signal intensity, and the signal-to-noise ratio, depend to some extent on the magnetic characteristics of the tissue being imaged. For a specific set of imaging factors, the tissue characteristics that enhance the signal-to-noise relationship are high magnetic nuclei (proton) concentration, short T1, and long T2. The primary limitation in imaging nuclei other than hydrogen (protons) is the low tissue concentration and the resulting low signal intensity.

TR and TE

Repetition time (TR) and echo time (TE) are the factors used to control contrast in most imaging methods. We have observed that these two factors also control signal intensity. This must be taken into consideration when selecting the factors for a specific imaging procedure.

When a short TR is used to obtain a T1-weighted image, the longitudinal magnetization does not have the opportunity to approach its maximum and produce high signal intensity. In this case, some signal strength must be sacrificed to gain a specific type of image contrast. Also, when TR is reduced to decrease image acquisition time, image noise can become the limiting factor.

When long TE values are used, the transverse magnetization and the resulting signal it produces can decay to very low values. This causes the images to display more noise.

RF Coils

The most direct control over the amount of noise energy picked from the patient's body is achieved by selecting appropriate characteristics of the RF receiver coil. In principle, noise is reduced by decreasing the amount of tissue within the sensitive region of the coil. Most imaging systems are equipped with interchangeable coils. These include a body coil, a head coil, and a set of surface coils as shown in Figure 10-4. The body coil is the largest coil and usually contains a major part of the patient's tissue within its sensitive region. Therefore, body

coils pick up the greatest amount of noise. Also, the distance between the coil and the tissue voxels is greater than in other types of coils. This reduces the intensity of the signals actually received by the coil. Because of this combination of low signal intensity and higher noise pickup, body coils generally produce a lower signal-to-noise ratio than the other coil types.

In comparison to body coils, head coils are both closer to the imaged tissue and generally contain a smaller total volume of tissue within their sensitive region. Because of the increased signal-to-noise characteristic of head coils, relatively small voxels can be used to obtain a better image detail.

The surface coil provides the highest signal-to-noise ratio of the three coil types. Because of its small size, it has a limited sensitive region

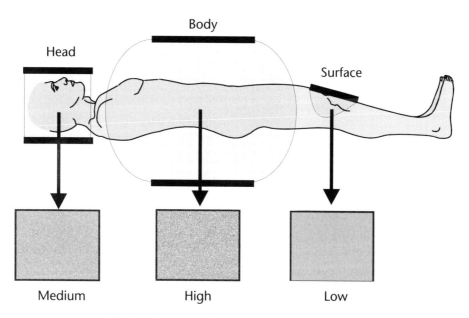

RADIO FREQUENCY COILS

Noise Received from Body

Figure 10-4. Both the amount of noise and the intensity of the signal received depend on the RF receiving coils. The body coil picks up the most noise and the weakest signal, resulting in the highest noise level in the image.

and picks up less noise from the tissue. When it is placed on or near the surface of the patient, it is usually quite close to the voxels and picks up a stronger signal than the other coil types. The compromise with surface coils is that their limited sensitive region restricts the useful FOV, and the sensitivity of the coil is not uniform within the imaged area. This non-uniformity results in very intense signals from tissue near the surface and a significant decrease in signal intensity with increasing depth. The relatively high signal-to-noise ratio obtained with surface coils can be traded for increased image detail by using smaller voxels.

Receiver Bandwidth

Bandwidth is the range of frequencies (RF) that the receiver is set to receive and is an adjustable protocol factor. It has a significant effect on the amount of noise picked up. This is because the noise is distributed over a wide range of frequencies, whereas the signal is confined to a relatively narrow frequency range. Therefore, when the bandwidth is increased, more noise enters the receiver. The obvious question is: Why increase bandwidth? One reason is that a wider bandwidth reduces the chemical shift artifact that will be described in Chapter 14. Also, wider bandwidths are the result of short signal sampling, or "picture snapping" times that are useful in some applications.

Averaging

One of the most direct methods used to control the signal-to-noise characteristics of MR images is the process of averaging two or more

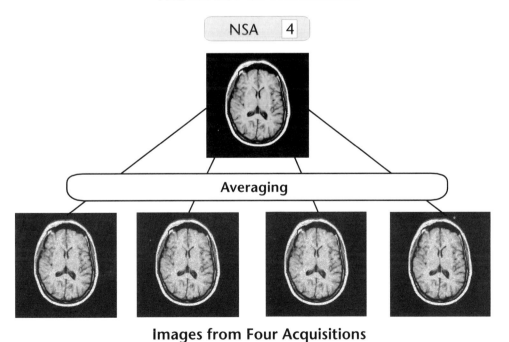

THE EFFECT OF AVERAGING

NSA 4

Averaging

Images from Four Acquisitions

Figure 10-5. An image with reduced noise is created by averaging the signals from four acquisitions.

signal acquisitions. In principle, each basic imaging cycle (phase-encoding step) is repeated several times and the resulting signals are averaged to form the final image as illustrated in Figure 10-6. The averaging process tends to reduce the noise level because of its statistical fluctuation nature, from one cycle to another. You can think of it as acquiring four images by repeating the basic acquisition four times. Then the signal intensities in each pixel position in the four images are averaged to produce an intensity value for the new averaged image.

The disadvantage of averaging is that it increases the total image acquisition time in proportion to the number of cycle repetitions or number of signals averaged (NSA). The NSA is one of the protocol factors set by the operator. Typical values are 1 (no averaging), 2, or 4, depending on the amount of noise reduction required. The general relationship is that the NSA must be increased by a factor of 4 to improve the signal-to-noise ratio by a factor of 2. The signal-to-noise ratio is proportional to the square root of the NSA. Sometimes the noise contribution from independent acquisitions adds; sometimes it cancels. Since the signals always add, adding or averaging independently acquired images improves the signal-to-noise ratio.

Mind Map Summary
Image Detail and Noise

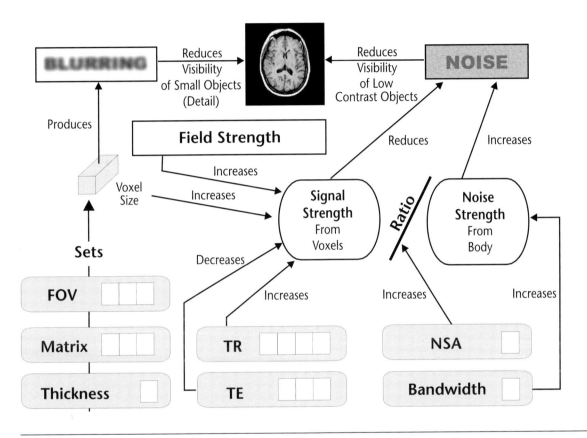

Two important image quality characteristics are blurring, which reduces visibility of small objects or detail, and image noise, which reduces visibility of low contrast objects. Both of these characteristics depend on design characteristics of the imaging system and the combination of selected protocol factors. The principal source of blurring in an MR image is the voxel size. This is because all tissues within an individual voxel are blurred together and represented by one signal. An image does not display any detail within the individual voxels. It is just a display of a matrix of voxels. Voxel size, and the resulting blurring, can be adjusted with the three protocol factors: FOV, matrix, and slice thickness.

The level of noise that appears in an image depends on the relationship (ratio) of the signal strength from the individual voxels and the noise strength coming from a region of the patient's body. The visible noise is reduced by increasing signal strength. This can be done by increasing the magnetic field strength, increasing voxel size, increasing TR, and decreasing TE. The field strength is a design characteristic and cannot be changed by the operator. Increasing voxel size to decrease noise has the adverse effect of also increasing blurring. Voxel sizes must be chosen to provide an appropriate balance between blurring and noise.

The noise strength picked up from the patient's body is determined by the mass of tissue contained within the sensitive pickup region of the RF coils. Surface coils that cover a relatively small anatomical region and are also close to the signal source (voxels) produce a high signal-to-noise relationship that results in lower image noise. The RF receiver bandwidth can be adjusted to block some of the noise energy from being received. However, decreasing the bandwidth to reduce noise has the adverse effect of increasing the chemical shift artifact.

Signal averaging is a useful technique for reducing noise but has the adverse effect of increasing acquisition time.

11

Acquisition Time And Procedure Optimization

Introduction And Overview

A significant time is usually required for the acquisition of an MR image or a set of multi-slice images. It is generally desirable to keep acquisition time as short as possible for a variety of reasons including: increasing patient throughput, reducing problems from patient motion, and performing dynamic imaging procedures. It is technically possible to reduce acquisition time by changing several protocol factors such as TR and matrix size. However, the basic protocol factors that affect acquisition time also have an effect on the characteristics and quality of the image. In general, changing factors to reduce acquisition time also decreases image quality.

When setting up an imaging protocol the acquisition time must be considered with respect to the necessary requirements for image quality. An optimized protocol is one in which there is a good balance among the image quality requirements and acquisition time.

In this chapter we will consider the various factors that determine acquisition time, how they relate to image characteristics and quality, and how the various requirements can be balanced.

Acquisition Time

Figure 11-1 illustrates a basic MR image acquisition process and identifies the factors that determine acquisition time. We recall that an image acquisition consists of an imaging cycle that is repeated many times. The duration of the cycle is TR, the protocol factor that is used to control certain image characteristics such as contrast. The primary factor that determines the number of times the cycle must be repeated is the number of rows of k space that must be filled. This, in turn, is determined by the image matrix size in the phase-encoded direction. As we observed in Chapter 9, it is the strength of the phase-encoding gradient during each cycle that directs the signals into a specific row of k space. Therefore, the phase-encoding gradient must be stepped through a range of different strengths, with the number of steps corresponding to the number of rows of k space that must be filled. The number of rows of k space

that must be filled is determined by the image matrix size in the phase-encoded direction. In Chapter 10 we saw that matrix size is one of the protocol factors that determines image detail and also has an effect on image noise. Reducing matrix size in an effort to reduce acquisition time without changing the FOV will decrease image detail.

In a basic acquisition the time required is TR multiplied by the matrix size in the phase-encoded direction.

$$\text{Time} = \text{TR} \times \text{matrix size}$$

In an acquisition, the time can be adjusted by changing either of these two factors. However, we will soon learn that there are some additional factors that are related to certain imaging

ACQUISITION TIME

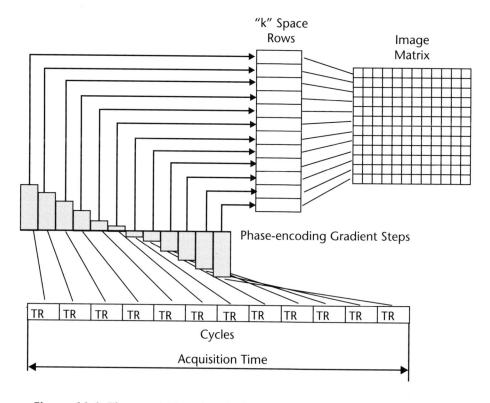

Figure 11-1. The acquisition time is determined by the cycle duration, TR, and number of cycles required to fill the k space.

techniques and methods that also have an effect on acquisition time. When applied, some will increase while others will decrease the time.

Cycle Repetition Time, TR

Until this point, we have used TR as an adjustable protocol factor to control the type of image (T1, T2, PD) that is being acquired. We recall from Chapter 6 that with spin echo imaging TR values of approximately 500 msec are used to acquire T1 images, but values as long as 2000 msec are required for PD and T2 images. TR is the time that is required, within an individual cycle, for the longitudinal magnetization to relax and recover to the appropriate level for the type of image that is being acquired. When the inversion recovery method is used, especially for fluid suppression (Chapter 8), even longer TR values are required. The point is that certain TR values are required to produce specific image types and they cannot be arbitrarily shortened to reduce acquisition time. As we learned in Chapter 7, much shorter TR values can be used with the gradient echo methods. This is why gradient

echo methods are generally much faster than the spin echo methods but are not appropriate for many imaging procedures.

Matrix Size

We recall from Chapter 10 that the matrix size and the FOV are the two factors that determine voxel size in the plane of the image. Voxel size determines the amount of blurring and image detail. Small voxels and minimum blurring are required for good detail. If the matrix size is reduced without changing the FOV, voxel size will be increased and there will be a reduction in image detail. Let us now use Figure 11-2 to see how matrix size can be adjusted to obtain an optimum balance between acquisition time and image detail.

It is only the matrix size in the phase-encoded direction that has an effect on acquisition time. Recall that this is because this matrix dimension determines the number of rows of k space that must be filled. Matrix size in the frequency-encoded direction does not affect acquisition time.

REDUCED ACQUISITION

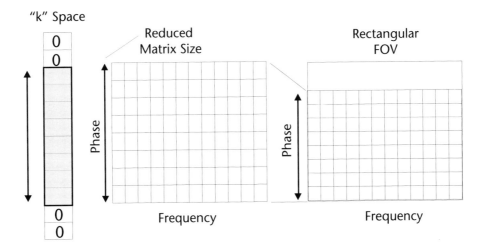

Figure 11-2. Reducing matrix size in the phase-encoded direction reduces the lines of k space to be filled. Then, reducing the FOV to a rectangle reduces the voxel size and restores image detail.

Reduced Matrix in Phase-Encoded Direction

One approach to optimizing an acquisition is to reduce the matrix size in the phase-encoded direction to a value that is less than the matrix size in the frequency-encoded direction. It is a common practice to set the matrix size in the phase-encoded direction to some percentage (such as 80%) of the size in the frequency-encoded direction. This has the effect of reducing acquisition time by 20% with minimal effect on image detail.

The selectable basic matrix sizes are binary multiplies such as 128, 256, 512, and 1024. This is necessary in order for the k space data to be in a form that can be easily processed in the image reconstruction phase. When basic matrix size is selected, it is one of these values, with 256 being the most common for most procedures. When the matrix size is reduced by some percentage in the phase-encoded direction, the computer fills in the unused row of k space with zeros to make the reconstruction process work.

Rectangular Field of View

Decreasing matrix size without changing the FOV does produce an increased voxel size. However, if the FOV can be reduced in the phase-encoded direction, the voxel size will be decreased and image detail will be maintained. The use of a rectangular FOV is a way of optimizing acquisition time and image detail if a rectangular FOV works for the specific anatomical region that is being imaged.

By combining a reduced matrix size (in the phase-encoded direction) with a rectangular FOV, acquisition time can be reduced without affecting image quality. This is one step in optimizing a procedure.

Half Acquisition

One way of filling k space with a reduced number of acquisition cycles is illustrated in Figure 11-3. In addition to half acquisition, this method might be referred to by names such as *half scan* and *half Fourier*.

This method is based on the fact that in a usual acquisition there is symmetry between the two halves of k space. This occurs because of the way the phase-encoding gradients are stepped as illustrated in Figure 11-1. Generally, during the first half of the acquisition the gradient is applied with a high strength for the first cycle. It is then stepped down from cycle to cycle reaching a very low, or zero, value near the center of the acquisition. During the second half of the acquisition the gradient strength is stepped back up to the maximum value for the last imaging cycle. In principle, the phase-encoding steps in the second half are a mirror image of the steps in the first half. Now let us return to Figure 11-3. When using the half acquisition method only the first half of k space is filled directly. Then, the data that was acquired during the first half is mathematically "flipped" and used to fill the second half of k space. This makes it possible to fill all of the rows of k space in approximately half of the normal acquisition time. The actual acquisition time will be slightly more than half because there must be some overlap in the data to make this process work. This is a method that can be used to reduce acquisition time. However, it results in an increase in image noise because the image is being formed with a smaller number of acquired signals. It has the opposite effect of using the technique of averaging to reduce noise that was introduced in Chapter 10. With respect to image noise, using half acquisition is equivalent to setting the NSA to a value of one half.

HALF ACQUISITION

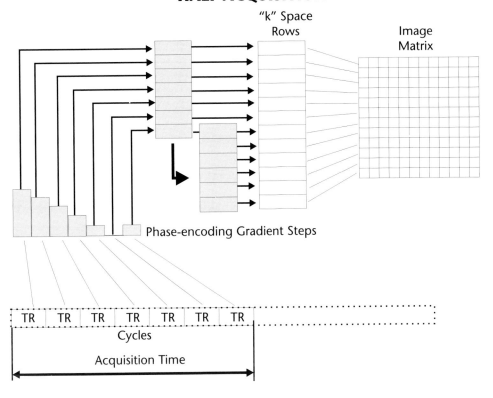

Figure 11-3. Acquisition time is reduced by using half acquisition and duplicating the data to fill the second half of k space.

Signal Averaging

We were introduced to the process of signal averaging in Chapter 10 as a technique that is used to reduce image noise. We now return to signal averaging to see how it affects acquisition time and how it can be combined with other factors to optimize an acquisition procedure. In Figure 11-4 we first see the general relationship between the image acquisition cycles and the phase-encoding gradient steps in a basic acquisition. This is an acquisition where neither signal averaging nor any of the fast imaging methods to be discussed later are used. In this basic acquisition one imaging cycle, with duration of TR, is used for each phase-encoding step. Therefore, the number of acquisition cycles is equal

to the number of lines in k space that must be filled, which is equal to the image matrix size in the phase-encoded direction. We will now see that there are acquisition techniques in which there is not this one-to-one relationship between the acquisition cycles and the phase-encoding steps.

The first of these is the use of signal averaging, which is also illustrated in Figure 11-4. The averaging is achieved by repeating the cycle several times for each phase-encoding step. This process does not change the number of phase-encoding steps required, but it does increase the number of cycles in the acquisition. In the example shown, the NSA protocol factor is set to 4. The cycle for each phase-encoding step is repeated four times and the total acquisition

SIGNAL AVERAGING

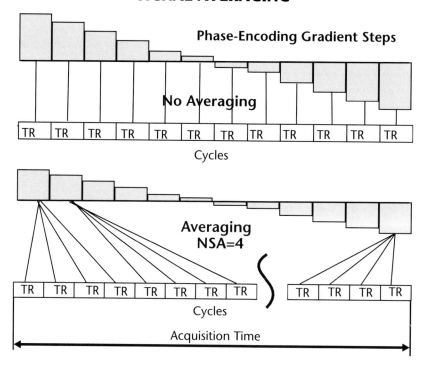

Figure 11-4. Acquisition time is increased by the factor NSA when averaging is used.

time is increased by a factor of 4. When averaging is used, the acquisition time becomes:

$$\text{Time} = \text{TR} \times \text{matrix size} \times \text{NSA}.$$

Protocol Factor Interactions

Up to this point we have observed that each of the protocol factors that affect acquisition time (TR, matrix size, and NSA) also have an effect on the image quality characteristics. The relationship becomes somewhat complex because some factors affect more than one characteristic. A good example is the factors that determine voxel size. As we have seen in Chapter 10 small voxels produce high image detail but result in higher image noise.

In Figure 11-5 we consider three important imaging goals that are affected by some of the same protocol factors. These goals are high

image detail, low image noise, and acquisition speed. We have placed these goals at the three corners of a triangle. We can think of this as a type of ball field with three goals. The objective is to move the ball closer to the goals. However, as we move the ball in the direction of one goal, we are moving away from the other goals. In this analogy the location of the ball represents the operating point of a specific acquisition protocol and is determined by the combination of protocol factors used. An optimized protocol is one in which the selected factors produce the best balance among high detail, low noise, and speed for a specific clinical procedure. In some procedures, high detail might be the most important characteristic, but it is obtained at some sacrifice of achieving low noise. The noise can then be reduced by averaging but at the cost of an increased acquisition time.

IMAGING GOALS

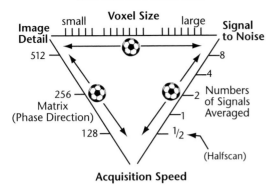

Figure 11-5. Several imaging goals must be considered together when selecting protocol factors.

Developing An Optimized Protocol

An acquisition protocol can be rather complex because of the large number of factors that must be adjusted and the interaction of many of the factors with different image quality characteristics and acquisition times. One approach to developing a good protocol is to address the specific image characteristics in this order:

- Contrast Sensitivity
- Image Detail
- Spatial Characteristics and Methods
- Image Noise
- Artifact Reduction

Contrast Sensitivity

The first step is to select an imaging method and factors (generally TR and TE) to give good contrast sensitivity and visualization for the specific tissue characteristic (T1, T2, PD) or type of fluid movement that is to be observed.

Also, consider if one of the selective signal suppression techniques (Chapter 8) will be helpful in improving contrast.

Image Detail

The next step is to consider the visibility of anatomical detail that is required for the specific clinical procedure. This should lead to the selection of an appropriate slice thickness, which is both the largest dimension of a voxel and the one that can be adjusted over a considerable range.

FOV is generally determined by the anatomy that is to be covered. A rectangular FOV can be used to permit reduced matrix in the phase-encoded direction without a reduction in detail.

Using a reduced matrix percentage in the phase-encoded direction can optimize the matrix size.

It is important to not go for more detail than is actually needed because, as we have seen, the small voxels cause an increase in image noise.

Spatial Characteristics and Methods

Generally, the 2-D multiple slice acquisition would be used unless thin slices are required for good detail and then the 3-D volume acquisition would be more appropriate.

The number of slices and slice orientations are determined by the anatomical regions that are to be covered and the desired views.

Image Noise

Voxel size (slice thickness, FOV, and matrix) has already been set to give the necessary image detail. TR and TE values have been selected to give the desired contrast characteristics. Also, magnetic field strength is a fixed value for the system being used. All these factors together, along with the anatomical volume of tissue that is to be imaged and the type of RF coil selected for the procedure, establish a basic level of image noise. This level of noise might or might not be acceptable. If it is not acceptable, averaging can be applied with the NSA parameter set to the minimum value that will give acceptable image quality, keeping in mind that this is increasing acquisition time.

Artifact Reduction

One of the final issues in developing a protocol is to consider the types of artifacts that might occur and then include in the protocol one or more of the artifact reduction techniques that are described in Chapter 14.

Fast Acquisition Methods

Until this point, we have seen that there are three factors that determine acquisition time: TR, matrix size, and NSA. This applies to all of the general methods in which only one line of k space is filled in each imaging cycle. We recall that this is when only one phase-encoding step occurs in each cycle. There are several methods that are capable of filling multiple rows of k space in one cycle. Two of these methods, echo planar and GRASE, were described in Chapter 7. Both of these are gradient echo methods. It is also possible to fill multiple rows of k space with the spin echo imaging methods by using the technique that we are about to describe. When this is done, the methods are generally known as *fast* or *turbo* spin echo.

There are many reasons to want to speed up an image acquisition process. It can increase patient throughput and reduce overall operating cost; shorter procedures are more comfortable for patients; there are less problems with patient motion. Also, some dynamic studies require the rapid acquisition of a series of images. Another advantage of using some of the fast acquisition methods is to compensate for other factors that increase acquisition time. A good example is when using a fluid suppression

technique as described in Chapter 8. This method requires very long TR values that would produce extremely long, and impractical, acquisition times. However, by combining it with a fast acquisition method, it becomes a very useful technique. The 3-D volume acquisition method that is used to produce thin slices and high image detail also requires a long acquisition time. This is because multiple repetitions must be made since phase-encoding is used to cut the slices, as described in Chapter 9. This becomes a more practical method when used with a fast acquisition technique.

Figure 11-6 illustrates the fast or turbo technique. We recall from Chapter 6 that it is possible to produce multiple spin echo signals within one cycle by applying several $180°$ pulses after each $90°$ excitation pulse. At that time we were using each of the echo signals to produce different images, each with a different TE value. This is the multiple echo method that makes it possible to produce both a PD image (with a short TE) and T2 images with the longer TE values in the same acquisition.

Fast or turbo spin echo uses multiple spin echo signals but for a different purpose. With this technique, a different phase-encoding gradient strength (step) is applied for each of the echo signals. Therefore, each echo signal fills a different row of k space, and multiple rows are filled within each cycle. The result is a reduced acquisition time. The reduction in time depends on the number of echo signals produced in each cycle and is one of the protocol variable factors. This number becomes the *speed factor* (or turbo factor) and acquisition time is now:

Time = TR × matrix size × NSA/speed factor.

Speed factors vary over a considerable range of possible values. In general, there is some

FAST SPIN ECHO

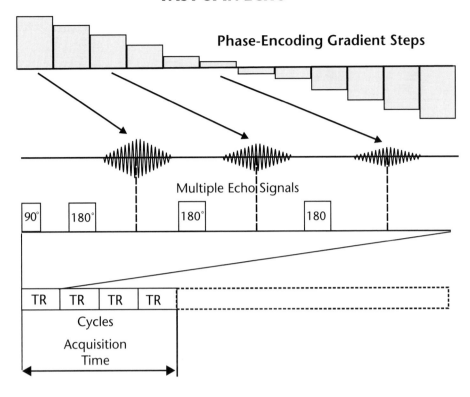

Figure 11-6. The fast spin echo method reduces acquisition time by producing multiple phase-encoded signals within each cycle.

reduction in signal strength (and increased noise) when high speed factor values are used.

With this technique k space is not filled in a progressive bottom-to-top order. The order of filling can be adjusted. One factor to consider is the TE value for the image. Since several echo signals, each with a different TE value, are produced in each cycle and used to form the same image, what is the "correct" TE for the image as far as contrast is concerned? When setting up the protocol for this method, an "effective" TE value is selected. This is the general TE value of the signals that are directed to the more central rows of k space. The data in these rows determine the general contrast characteristics of the image. The data in the outer rows of k space contribute more to the detail characteristics of the image.

Mind Map Summary
Acquisition Time And Procedure Optimization

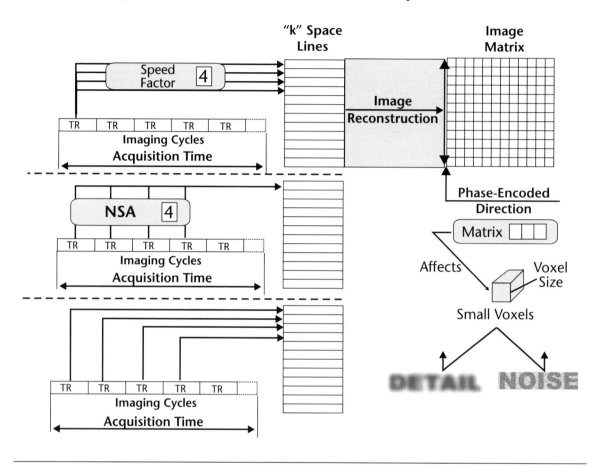

The MR image acquisition process consists of an imaging cycle (like a heartbeat) that must be repeated many times. The total acquisition time is determined by the duration of each cycle; that is the protocol factor, TR, and the number of cycles required. In a basic imaging procedure, the number of cycles is determined by the number of rows of k space that must be filled, which, in turn, is determined by the image matrix size in the phase-encoded direction. The matrix size can be decreased but this results in an increased voxel size and reduced image detail. When setting up an imaging protocol, attention should be given to selecting a voxel size that provides an appropriate balance between image detail and image noise.

Signal averaging is used to reduce image noise, but it also results in an increased acquisition time because cycles must be repeated according to the selected NSA value.

There are several imaging methods that are capable of filling multiple lines of k space during one imaging cycle. These are some of the fast imaging methods that include fast (turbo) spin echo, echo planar, and GRASE. There is an adjustable speed factor associated with each of these methods.

12

Vascular Imaging

Introduction And Overview

One of the important characteristics of MRI is its ability to create an image of flowing blood and vascular structures without having to inject contrast media. With MRI the contrast between the blood and the adjacent stationary tissue is produced by interactions between the *movement* of the blood and certain events within the imaging process. This is very different from x-ray angiography where the image shows the presence of blood (or contrast media). In MRI the image displays blood-filled vessels, but it is the movement of the flood that actually produces the contrast.

It is a somewhat complex process because under some imaging conditions the flowing blood will produce increased signal intensity and will appear bright, while under other conditions very little or no signal will be produced by the flowing blood and it will be dark. We will designate these two possibilities as "bright blood" imaging and "black blood" imaging, as indicated in Figure 12-1. There are several different physical effects that can produce both bright blood and black blood as indicated. These effects can be divided into three categories:

1. Time effects.
2. Selective saturation.
3. Phase effects.

Under each of these categories there are several specific effects that can produce contrast. Unfortunately, in addition to producing

FLOW EFFECTS

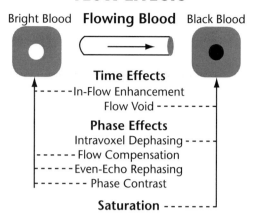

Figure 12-1. The physical effects that produce contrast between flowing blood and the surrounding tissue background.

FLOW ENHANCEMENTS

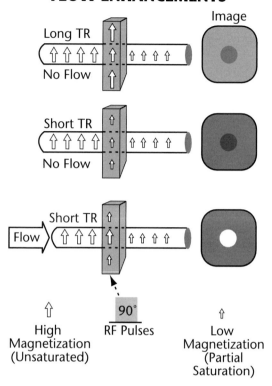

Figure 12-2. Flow-related enhancement produces bright blood because the background tissue is partially saturated by the use of short TR values.

useful image contrast, the movement of blood often produces undesirable artifacts within the image. These will be considered in Chapter 14.

Some of the flow effects can appear in virtually any image that contains blood vessels. In this chapter we will first explore the various effects associated with flowing blood and show how they can be controlled and used. Then we will apply these effects to specific angiographic procedures.

Time Effects

The time effects are related to the movement of blood during certain time intervals within the acquisition cycle. They are sometimes referred to as the *time-of-flight* effects. The production of bright blood is related to the TR time interval whereas the production of black blood is related to the TE time interval.

Flow-Related Enhancement (Bright Blood)

The process that causes flowing blood to show an increased intensity, or brightness, is illustrated in Figure 12-2. This occurs when the

direction of flow is through the slice, as illustrated. It is also known as the *in-flow* effect and is used to produce contrast in in-flow contrast angiography. The degree of enhancement is determined by the relationship of flow velocity to TR. Three conditions are illustrated. The arrow indicates the amount of longitudinal magnetization at the end of each imaging cycle. Because of the slice-selection gradient the RF pulses affect only the blood within the slice.

Let us start by considering using a long TR in the absence of flow in the vessel. The longitudinal magnetization regrows to a relatively high value during each cycle, as indicated at the top of the figure. This condition produces a relatively bright image of both the blood and the stationary tissue and little contrast between the two. With this as a reference condition, let us now see what happens when a short TR value is used. Each cycle will begin before the longitudinal magnetization has approached its maximum. This results in reduced signal intensity and a relatively dark image because both the non-flowing blood and tissue remain partially saturated.

The next step is to see what happens if the blood is flowing into the slice. The effect of flow is to replace some of the blood in the slice with fully magnetized blood from outside the slice. The increased magnetization at the end of each cycle increases image brightness of the flowing blood. The enhancement increases with flow until the flow velocity becomes equal to the slice thickness divided by TR. This represents full replacement and maximum enhancement and brightness.

In order for blood to appear bright, the surrounding background stationary tissue must be relatively dark. Therefore, a part of any bright blood imaging process is to suppress the signals from the surrounding tissue. With the in-flow effect, the background tissue is dark because it is partially saturated from the use of the short TR values. When long TR values are used, the background tissue is not suppressed and there is little, if any, contrast with the flowing blood.

There are several factors that can have an effect on flow enhancement. In multi-slice imaging, including volume acquisition, the degree of enhancement can vary with slice position. Only the first slice in the direction of flow receives fully magnetized blood. As the blood reaches the deeper slices, its magnetization and resulting signal intensity will be reduced by the RF pulses applied to the outer slices. Slowly flowing blood will be affected the most by this. Faster flowing blood can penetrate more slices before losing its magnetization. Related to this

is a change in the apparent cross-sectional area of enhancement from slice to slice. A first slice might show enhancement for the entire cross section of a vessel. However, when laminar flow is present, the deeper slices will show enhancement only for the smaller area of fast flow along the central axis of the vessel. Any other effects that produce black blood can counteract flow-related enhancement. One of the most significant is the flow-void effect, which takes over at higher flow velocities.

Bright blood from flow-related enhancement is especially prevalent with gradient-echo imaging. There are two major reasons for this. The short TR values typically used in SAGE imaging increase the flow-related enhancement effect. Also, when a gradient rather than an RF pulse is used to produce the echo, there is no flow-void effect to cancel the enhancement. Therefore, gradient echo imaging is used when using the in-flow process to produce bright blood specifically, as in MR angiography.

Flow-Void Effect (Black Blood)

Relatively high flow velocities through a slice reduce signal intensity and blood brightness with spin echo imaging. Figure 12-3 illustrates this effect. The flow-void effect occurs because the blood flows out of the slice between the 90° and the 180° pulses. The arrow in the slice indicates the level of residual transverse magnetization that is present when the 180° pulse is applied to form the spin echo signal. This is the transverse magnetization produced by the preceding 90° excitation pulse. The time interval between the 90° and the 180° pulses is one-half TE. If the blood is not moving, the blood that was excited by the 90° pulse will be within the slice when the 180° pulse is applied. This results in maximum rephasing of the transverse protons at the echo event and relatively bright blood.

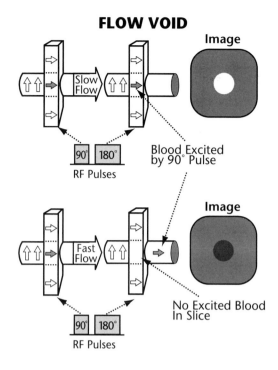

FLOW VOID

Figure 12-3. The flow-void effect caused by blood flowing out of the slice between the 90° and 180° pulses.

If the blood moves out of the slice between the 90° and 180° pulses, complete rephasing and the formation of an echo signal will not occur. This is because the 180° pulse can affect only the blood within the slice. The spin echo signal is reduced, and the flowing blood appears darker than blood moving with a lower velocity. The intensity continues to drop as flow is increased until the flow velocity removes all magnetized blood from the slice during the interval between the two pulses (one-half TE).

Selective Saturation

As we have observed on several occasions, saturation of the longitudinal magnetization with a 90° RF pulse produces a temporary darkness of a tissue or fluid because there is little magnetization to produce a signal. We saw this

applied in Chapter 8 as a method for reducing artifacts from moving tissue. Selective saturation of a specific anatomical region is an effective way of producing black blood.

This technique is generally known as presaturation because the saturation pulse is applied before the imaging pulses in each cycle, and is applied to the blood before it enters the image slice. Figure 12-4 illustrates this concept. An RF pulse is selectively applied to the anatomical region that supplies blood to the slice. This pulse destroys the longitudinal magnetization by flipping it into the transverse plane. When this saturated, or unmagnetized, blood flows into the slice a short time later, it is not capable of producing a signal. Therefore, the image displays a void or black blood in the vessels. The region, or slab, of presaturation can be placed on either side of the imaged slice. This makes it possible to selectively turn off the signals from blood flowing in opposite directions.

The presaturation technique can be used to: (1) produce black-blood images; (2) selectively image either arterial or venous flow; and (3) reduce flow-related artifacts, as described in Chapter 14.

Phase Effects

There are two important phase relationships that can be affected by movement of blood during the imaging process. One is the phase relationship among the spinning protons within each individual voxel (intravoxel) and the other is the voxel-to-voxel (intervoxel) relationship of the transverse magnetizations, as shown in Figure 12-5. Both of these concepts have been described before. We will now see how they are affected by flowing blood, how they contribute to image contrast, and how they can be compensated for by the technique of flow compensation.

SATURATION

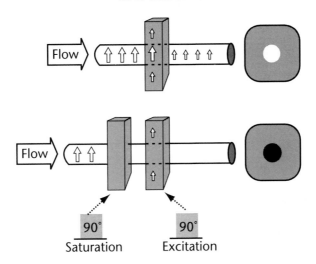

Figure 12-4. The application of a saturation pulse to eliminate the signals from flowing blood.

PHASE EFFECTS OF FLOW

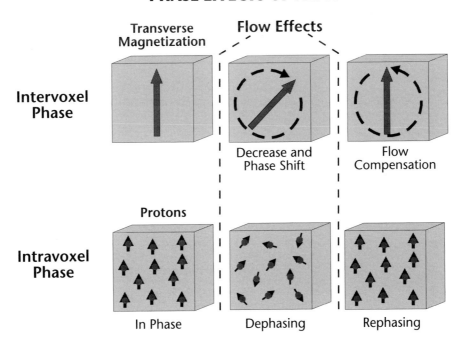

Figure 12-5. Changes in both intervoxel and intravoxel phases produced by flow.

Intravoxel Phase

In order to produce a signal, the protons within an individual voxel must be in-phase at the time of the echo event. In general, dephasing and the loss of transverse magnetization occur when the magnetic field is not perfectly homogeneous or uniform throughout a voxel. A gradient in the magnetic field is one form of inhomogeneity that produces proton dephasing. Since gradients are used for various purposes during an image acquisition cycle, this dephasing effect must be taken into account. For stationary (non-flowing) tissue or fluid the protons can be rephased by applying a gradient in the opposite direction, as shown in Figure 12-6. We saw how a gradient was used to diphase and then rephase protons to produce a gradient echo event in Chapter 7. Let us now consider this process in more detail and apply it to flowing blood.

In general, we are considering events that happen within the TE interval; that is, between the excitation pulse and the echo event. We recall that the phase-encoding gradient is applied during this time. We will use it as the example for this discussion. When the gradient is applied as shown, the right side of the voxel is in a stronger magnetic field than the left. This means that the protons on the right are spinning faster than those on the left and quickly get out of phase. However, they can be rephased by applying a gradient in the reverse direction as shown. Now the protons on the left are in the stronger field and will be spinning faster. They will catch up with and come into phase with the slower spinning protons on the right. At the time of the echo event the protons are in-phase and a signal is produced. The process just described is used in virtually all imaging methods to compensate for gradient dephasing.

PHASE CHANGES AND FLOW COMPENSATION

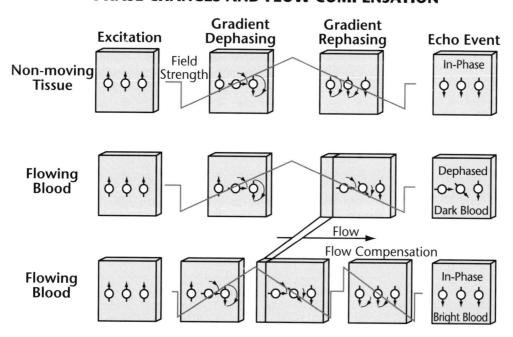

Figure 12-6. Dephasing and rephasing produced by gradients and flow compensation.

Flow Dephasing

Our next step is to consider what happens to the protons in a voxel of flowing blood or other fluid. This is also illustrated in Figure 12-6. If the voxel is moving, the protons will not be completely rephased by the second gradient. This is because the protons will not be in the same position relative to the gradient and will not receive complete compensation. The result of this is that flow in the direction of a gradient will generally produce proton dephasing and little or no signal at the time of the echo event. This is one possible source of black blood.

Flow Compensation

There are techniques that can be used to rephase the protons within a voxel of flowing blood or other fluid. These techniques use a somewhat complex sequence of gradients to produce the rephasing, as shown in Figure 12-6. The technique is generally called *flow compensation* or *gradient moment nulling.*

One of the selectable protocol factors controls the characteristics of the flow compensation gradients. The technique can be turned on or off and adjusted to compensate for specific types of flow. Flow with a constant velocity is the easiest to compensate. It is possible to compensate pulsatile (changing velocity or acceleration) flow by using a more complex gradient waveform. Flow compensation is somewhat velocity dependent. That is, all velocities are not equally compensated by the same gradient waveform.

One factor that must be considered is the time required within the TE interval for the flow compensation process. Some of the more complex compensation techniques might increase the shortest TE that can be selected. Flow compensation gradients can also be used to compensate for flow-induced *intervoxel* phase changes, which are described later.

There are two major applications of flow compensation: One is to reduce flow-related artifacts (Chapter 14) and loss of signal; and the other is as a part of phase contrast angiography, which will be developed later in this chapter.

Even-Echo Rephasing

The phenomenon of even-echo rephasing can be observed when a multi-spin echo technique is used to image blood flowing at a relatively slow and constant velocity. This effect produces an increase in signal intensity (brightness) in the even-echo images (second, fourth, etc.) compared to the odd-echo images.

This effect occurs when there is laminar flow through the vessel and the individual voxels. This means that the different layers of protons within a voxel are moving at different velocities and will experience different phase shifts as they flow through a gradient. This is a dephasing effect that will reduce transverse magnetization and signal intensity at the time of the first echo event. The key to even-echo rephasing is that the second 180° pulse reverses the direction of proton spin. Before the pulse, the protons in the faster layers had moved ahead and gained phase on the slower moving protons. However, immediately after the pulse they are flipped so that they are behind the slower-spinning protons. This sets the stage for them to catch up and come back into phase. This occurs at the time of the second echo event and results in an increase in transverse magnetization, higher signal intensity, and brighter blood than was observed in the first echo image.

The alternating of blood brightness between the odd- and even-echo images will continue for subsequent echoes, but there will be an overall decrease in intensity because of the T2 decay of the transverse magnetization. Both turbulent and pulsatile flow tend to increase proton dephasing within a voxel, which results in a loss of signal intensity. Under some conditions this can counteract the effect of even-echo rephasing.

Intervoxel Phase

If a voxel of tissue moves during the image acquisition process, the phase relationship of its spinning transverse magnetization can be shifted relative to that of other voxels. There are both advantages and disadvantages of this effect.

Phase Imaging

Most MRI systems are capable of producing phase images. The phase imaging that is described here is different from phase contrast angiography that will be described later. A phase image is produced with one of the conventional imaging methods but with a different way of calculating the image from the signals. A phase image is one in which the brightness of a voxel is determined by its phase relationship rather than the magnitude of transverse magnetization. This is illustrated in Figure 12-7.

Consider a row of voxels across a vessel through which blood is flowing. We will assume laminar flow with the highest velocity along the central axis of the vessel. As this blood flows through a magnetic field gradient, the phase of the individual voxels will shift in proportion to the flow velocity. There, if we create an image at a specific time, we can observe the phase relationships. If the phase of a voxel's transverse magnetization is then translated into image brightness (or perhaps color), we will have an image that displays flow velocity and direction. This type of image is somewhat analogous to a Doppler ultrasound image in that it is the velocity that is being measured and displayed in the image.

Analytical software can be used in conjunction with phase images to calculate selected flow parameters.

PHASE IMAGE PIXEL BRIGHTNESS

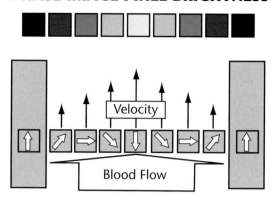

Figure 12-7. Intervoxel phase changes can be used to measure and image blood flow.

Artifacts

Flow and other forms of motion can produce serious artifacts in MR images. Although the techniques that can be used to suppress artifacts will be described in Chapter 14, it is appropriate to consider the source of one such artifact here. We recall that the process of phase-encoding is used to produce one dimension in the MR image. A gradient is used to give each voxel in the phase-encoded direction a different phase value. This phase value of each voxel is measured by the reconstruction process (Fourier transform) and used to direct the signals to the appropriate image pixels, as described in Chapter 9. This process works quite well if the tissue voxels are not moving during the acquisition process. However, if a voxel moves through a gradient, the phase relationship of its transverse magnetization will be altered, as described above. This means that the RF signal will no longer carry the correct phase address and will be directed to and displayed in the wrong pixel location. The observable effect is streaking or ghost images in the phase-encoded direction. Although any type of tissue motion can produce this type of artifact, it is a significant problem with flowing blood. Flow compensation as described above, and other techniques to be described in Chapter 14 are used to reduce this type of artifact.

Angiography

The MR angiography methods use a combination of the effects described above to produce vascular images. Generally, MR angiography is used to produce vascular images covering a thick anatomical volume as opposed to a relatively thin slice as in most other imaging methods. There are several different approaches used to produce MR angiograms. Each method has specific characteristics that must be considered when producing clinical images. The three general methods based on how the contrast between blood and background tissue are produced are:

1. Phase Contrast Angiography.
2. In-Flow Contrast Angiography.
3. Contrast Enhanced Angiography.

The first two methods derive contrast from the movement of the blood and the flow effects that have just been described. The third method uses administered contrast media to enhance the vascular contrast.

Phase Contrast Angiography

We recall that when blood flows through a magnetic-field gradient, the phase relationship is affected. This applies both to the phase relationship of protons within a voxel and the voxel-to-voxel relationship of transverse magnetization. Both of these—intravoxel and intervoxel—phase effects can be used to produce contrast of flowing blood. However, the intervoxel phase shift of the transverse magnetization is the effect most frequently used to produce phase contrast angiograms.

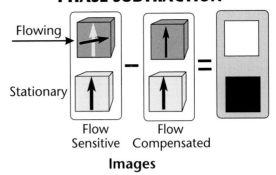

PHASE SUBTRACTION

Figure 12-8. The principle of phase contrast angiography. Blood brightness is related to the phase change produced by flow.

When blood flows through a gradient, the phase of the transverse magnetization changes in proportion to the velocity. When this technique is used, the phase shift, not the magnitude of the magnetization, is used to create the image. Therefore, blood brightness is directly related to flow velocity in the direction of the flow-encoding gradient.

Figure 12-8 illustrates the basic process of creating a phase contrast angiogram. At least two image acquisitions are required. One image is acquired with a flow phase-encoding gradient turned on. The phase of the magnetization is shifted in proportion to the flow velocity. A flow compensation gradient is applied during the acquisition of a second image to reset the phase of the flowing blood. The phase in the stationary tissue is not affected and is the same in both images.

The mathematical process of phase or vectory subtraction used to produce the phase contrast image is illustrated in Figure 12-9. The vector subtraction process determines the difference in phase between the flow-encoded blood and the flow-compensated blood, which serves as a reference. The phase difference (and

PHASE SUBTRACTION

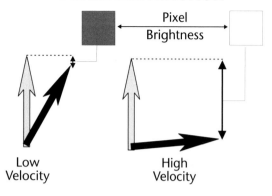

Figure 12-9. The principle of phase subtraction.

blood brightness) increases with velocity. However, when the phase reaches the 6 o'clock position, additional increases in velocity will produce a reduced difference in phase and brightness as the phase returns to the 12 o'clock reference position. The contrast in this type of image is related to both the velocity and direction of the flowing blood.

Velocity

Because the degree of phase shift and the resulting contrast is related to flow velocity in a cycling manner, the operator must set a velocity value as one of the protocol factors. This will give a specific relationship between phase values and velocity for that particular imaging procedure. Flow at this rate will produce maximum contrast as shown in Figure 12-10. The important thing to note is that at some velocities, both below and above the set velocity, the blood will be black. This can produce an aliasing artifact in which fast-flowing blood will be dark and will look like, or alias, slow-flowing blood. This must be taken into account both when selecting a set velocity value and when viewing images. An advantage of phase contrast angiography is the ability to image specific ranges of flow, including relatively slow flow rates if the proper factors are used.

Flow Direction

Phase contrast is produced only when the flow is in the direction of a gradient. To image

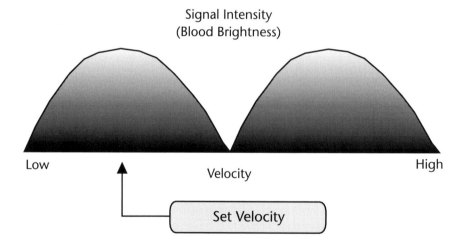

Figure 12-10. The relationship of blood brightness to flow velocity in phase contrast angiography.

blood that is flowing in different directions, several image data sets must be acquired with gradients in the appropriate directions. Flow in all possible directions can be imaged by acquiring three images with three orthogonal gradient directions.

The images for the different flow directions are combined with the subtraction process to produce one composite angiographic image.

In-Flow Contrast Angiography

With the in-flow contrast method an angiographic image is created by using the flow-related enhancement principle described previously. This is also called the time-of-flight technique. Bright blood images are produced by using a gradient echo method with relatively short TR values. Because of the short TR values, the background tissue remains partially saturated (and dark) while the flowing blood produces a stronger (bright) signal because of the in-flow effect.

To produce an angiogram, the method must be extended to cover an anatomical volume rather than a single slice. There are two acquisition techniques that can be used to achieve this: 3-D volume acquisition or multiple 2-D slice acquisitions. Each has its special characteristics that must be considered in clinical applications.

Three-Dimension (3-D) Volume Acquisition

A 3-D volume acquisition can be used to create an angiographic image, as shown in Figure 12-11. We recall (from Chapter 9) that with this acquisition method RF pulses are applied to and signals are acquired from an entire volume simultaneously. During acquisition a phase-encoding gradient is applied in the slice selection direction. The individual slices are then created during the image reconstruction process. The principal advantage is the ability to create thin, contiguous slices with small voxel dimensions. This gives good image

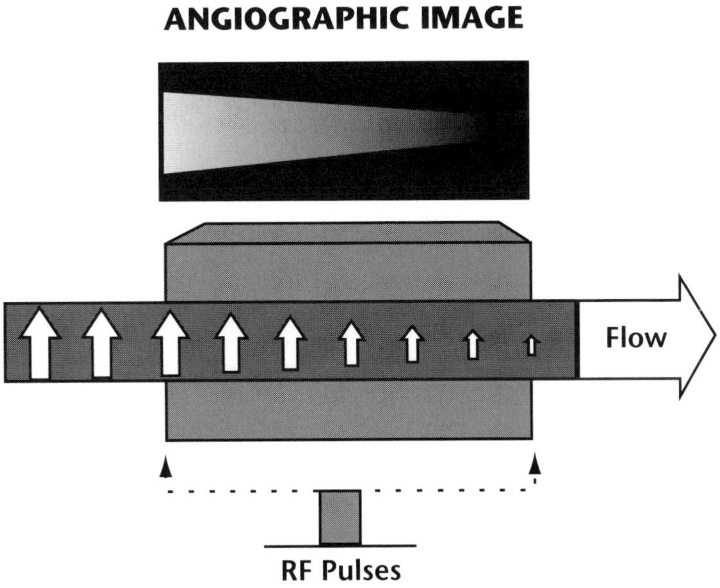

ANGIOGRAPHIC IMAGE

Figure 12-11. A 3-D volume acquisition showing decreased magnetization and blood brightness produced by saturation as the blood flows deeper into the volume.

detail. Also, in vascular imaging small voxels reduce intravoxel dephasing and signal loss.

The principal disadvantage of this method is that relatively slow-flowing blood becomes saturated as it passes through the acquisition volume. This can limit the volume size and vessel length, which is capable of producing good image contrast. Blood flowing with a relatively high velocity, as along the central axis of a vessel, will be imaged deeper into the acquisition volume than slow-flowing blood. This is illustrated in Figure 12-11.

Two-Dimension (2-D) Slice Acquisition

With this technique the anatomical volume is covered by acquiring a series of single-slice images, as shown in Figure 12-12. This approach provides relatively strong signals (bright blood) throughout the volume because each slice is imaged independently and not affected by blood saturated in other slices. This makes it possible to image relatively long vessels extending over large volumes. Also, with this method, the signals from the flowing blood are relatively independent of flow velocity.

Image Format

The image acquisition and reconstruction process results in data in the form of many slice images covering a volume within the patient's body. This must then be reformatted into a 3-D image of the vascular structure extending over this volume. There are several methods that can be used for this purpose.

Maximum Intensity Projection (MIP)

The maximum intensity projection (MIP) technique is commonly used to create a single com-

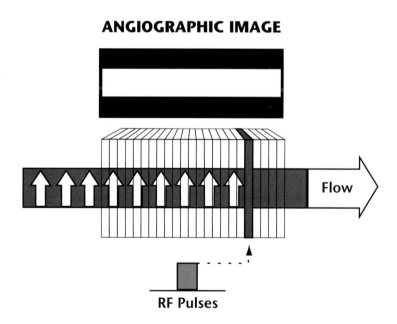

ANGIOGRAPHIC IMAGE

Flow

RF Pulses

Figure 12-12. The acquisition of thin slices reduces the saturation and loss of signal intensity within the imaged volume. Compare to figure 12-11.

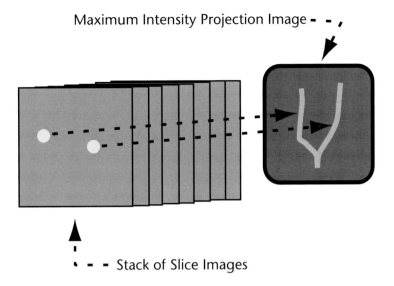

Maximum Intensity Projection Image

Stack of Slice Images

Figure 12-13. The use of maximum intensity projection (MIP) to produce a composite image from a stack of slice images.

posite image from a stack of images, as shown in Figure 12-13. This is a mathematical process performed by the computer. The stack of images is "viewed" along a series of pathways through the volume. The maximum signal intensity, or blood brightness, encountered in any slice along each pathway is then projected onto the composite image. The resulting MIP image is a 2-D image of the 3-D vascular structure. This process generally enhances the contrast between the flowing blood and the stationary tissues.

It is possible to create images by projecting in different directions through the volume. This gives the impression of rotating the vascular structure so that it can be viewed from different directions.

Surface Rendering

Images can also be formed by using surface rendering programs. This is especially useful for large vessels such as the aorta.

Contrast Enhanced Angiography

Images of vascular structures can be produced by injecting contrast media and not relying on the time and phase effects described previously to produce the image contrast. This overcomes some of the problems of signal loss produced by saturation in large areas and the limited velocity range than can be imaged with phase contrast. It is, however, restricted in terms of depending on the presence of contrast media in the vessels for continuous imaging.

Mind Map Summary
Vascular Imaging

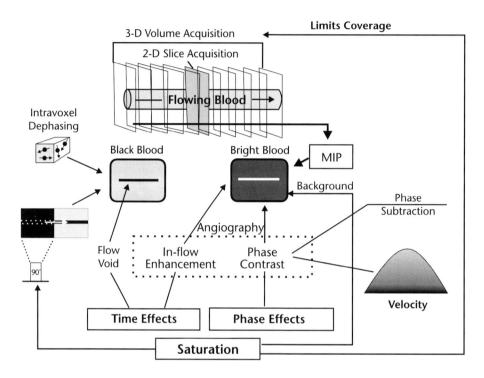

MR is capable of producing images of flowing blood without the injection of contrast media. The visibility or contrast of the blood is produced by a variety of physical effects or interactions between the moving blood and the MR process. Some effects produce black blood and others produce bright blood. Black blood occurs when the flowing blood does not produce a signal. This can be caused by the flow void effect, the application of saturation pulses to a vascular area, or any condition that produces intravoxel dephasing.

Bright blood effects, used in angiography, are in-flow enhancement and phase contrast. In-flow enhancement produces bright blood by keeping the surrounding stationary tissue partially saturated. Blood flowing into the image area is not saturated and produces a bright signal. However, consideration must be given to how the image slices are formed and acquired over an anatomical region. 3-D volume acquisition produces thin slices and high image detail, but results in partial saturation of the flowing blood and possible deterioration of the image. An alternative method that produces less saturation of the flowing blood is the 2-D slice acquisition process.

In phase contrast angiography the transverse magnetization of flowing blood experiences a phase shift when passing through a gradient. This phase change, which is proportional to velocity, can be measured and translated into brightness in an image. Most MR angiographic images represent a 3-D anatomical region and are produced from a stack of slice images by the MIP process or by surface or volume rendering image processing.

13

Functional Imaging

Introduction And Overview

MR functional imaging consists of several different imaging methods that are used to visualize and, in some cases, quantify blood and fluid movement beyond the general vascular system. In the last chapter we learned that there are several techniques that can be used to produce contrast and a visible image of blood flowing through vessels. It is the perfusion throughout the capillary bed and then the diffusion of fluids throughout the tissue that is the subject of most MR functional imaging procedures. Specific and different methods are used to image both the perfusion and diffusion processes and to visualize tissue areas that are metabolically active.

Diffusion Imaging

Diffusion imaging is generally used to visualize tissue areas in which some pathologic process has altered the motion of fluid that is outside of the vessels and capillaries. This is the natural, random, incoherent, Brownian motion of the proton-containing molecules. The motion is always there, but its rate is what changes and is the source of contrast in diffusion imaging.

Brownian Motion and Diffusion

Free or unbound molecules in a material such as fluid are in constant motion because of thermal activity. This is a random motion in which the molecules travel in one direction, collide with other molecules and change direction,

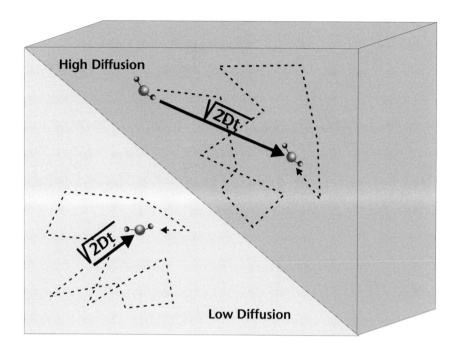

Figure 13.1. The average distance traveled by a molecule in time, t, depends on the diffusion coefficient (D) for the specific tissue compartment within a voxel.

and then move in another direction until the next collision. This is known as Brownian motion and is illustrated in Figure 13-1. It is this motion that is the source of the general process of diffusion. Diffusion is what accounts for the net movement of a substance from an area of high concentration to an area of lower concentration in the absence of other pressures or forces. Within a region with a uniform concentration, such as a voxel, the molecules are still in motion even if there is no net movement from one region to another.

The distance that the molecules move in a specific time is very dependent on structural characteristics of this tissue. The motion is generally greatest in fluids and somewhat less in a cellular tissue environment.

Diffusion Coefficient

The diffusion coefficient, D, is related to the net displacement of molecules in a given time. The net displacement distance is much less than the total path length traveled by a molecule during this time. This is a statical process in which there is a range of distances traveled by molecules in the same period of time. The mean (rms) distance traveled is what is related to the diffusion coefficient as shown in Figure 13-1.

Apparent Diffusion Coefficient (ADC)

Diffusion coefficient values determined by MRI might be a composite from several structural compartments (extracellular and intracellular) within a voxel. There could be different diffusion coefficient values within these various compartments. Therefore, the values determined are designated as the apparent diffusion coefficient, ADC.

Diffusion Direction

In a specific tissue the diffusion rate might be different in different directions because of the orientation of certain tissue structures. This is an important factor that must be taken into account when producing diffusion images. As we will soon discover, the sensitivity for observing diffusion is produced by applying magnetic field gradients. Only when a gradient is applied in the direction of the diffusion will it be imaged.

The Imaging Process

Diffusion imaging is based on the principle that the diffusion motion of the molecules produces a dephasing of the spinning protons within a voxel and that this results in a reduced signal intensity and image brightness. The dephasing is actually produced by applying additional diffusion-sensitizing gradients during the image acquisition cycle as shown in Figure 13-2. Here we see the gradients used in conjunction with a spin echo pulse sequence. Notice that there are actually two gradient pulses. One is applied before the 180° RF pulse and the second gradient is applied after the pulse. Let us recall that a gradient is a variation in field strength across each individual voxel. During the time of the gradient the spinning protons will be in different field strengths and

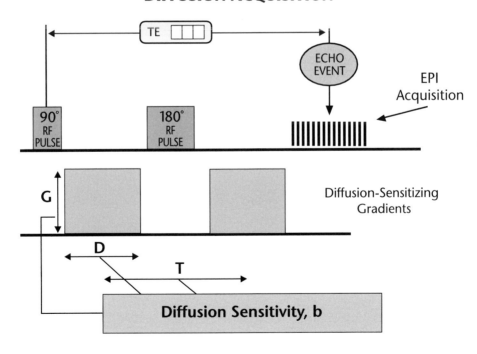

Figure 13.2. The application of gradients to produce dephasing and reduced signal intensity related to diffusion.

spinning at different rates, along the direction of the gradient. This produces a dephasing of the protons within the voxel. When the 180° RF pulse is applied, it reverses the spin direction. Now when the second gradient is applied, it produces a rephasing on the spinning protons within the voxel. However, only the protons that have not moved or changed position between the times of the two gradients will be completely rephased. The protons in molecules that have moved will be in a different location and field strength during the second gradient and will not be completely rephased. This results in reduced signal intensity from voxels containing the moving protons. It is this reduction in signal intensity that produces the contrast of the diffusing molecules with respect to the non-moving tissue structures.

The reduction in signal intensity (S_D/S_0) produced by the diffusion depends on two factors: the rate of diffusion expressed by the value of the diffusion coefficient, D, and the diffusion sensitivity, b, which is determined by characteristics of the gradients. The relationship is given by:

$$S_D/S_0 = e^{-bD}$$

Diffusion Sensitivity

The diffusion sensitivity, b, is determined by the strength, duration, and time separating the

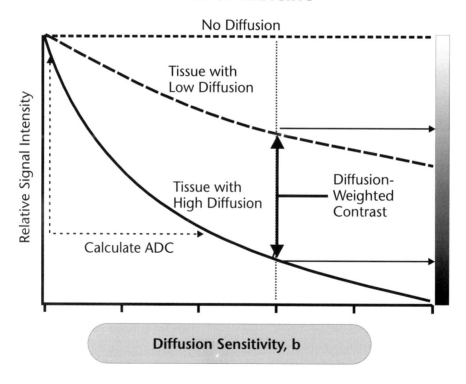

DIFFUSION IMAGING

Figure 13-3. The relationship of signal intensity to the diffusion rates in tissues and the setting of the diffusion sensitivity protocol parameter.

two gradients. These factors are identified in Figure 13-2. The sensitivity is very dependent on the gradient strength, G. Maximum gradient strength is a design characteristic of the imaging system. Diffusion imaging is generally limited to systems equipped with strong, high-performance gradients.

The value of the diffusion sensitivity, b, is an adjustable protocol factor. Images can be acquired with different sensitivity values.

Diffusion-Weighted Images

Images in which the diffusion rate is a source of contrast can be obtained by applying the diffusion-sensitizing gradients as described. The sensitivity can be adjusted by changing the b protocol factor. As the value of b is increased, diffusion has an increased weighting and areas with relatively high levels of diffusion become darker. This is illustrated in Figure 13-3.

ADC Map Images

An alternative to diffusion-weighted images is to calculate the apparent diffusion coefficient, ADC, for each voxel and display the values in the form of an image. This can be done because the slope of the signal intensity versus b value curve is determined by the ADC of each specific tissue voxel. By acquiring images at several (at least two) b values, the ADC can be calculated. ADC map images have contrast that is opposite to diffusion-weighted images. Areas of increased diffusion are bright.

Blood Oxygenation Level Dependent (BOLD) Contrast

Blood oxygenation level dependent, or BOLD, is a source of contrast related to neural activity that is endogenous and does not require the administration of any contrast agent. Advantages are

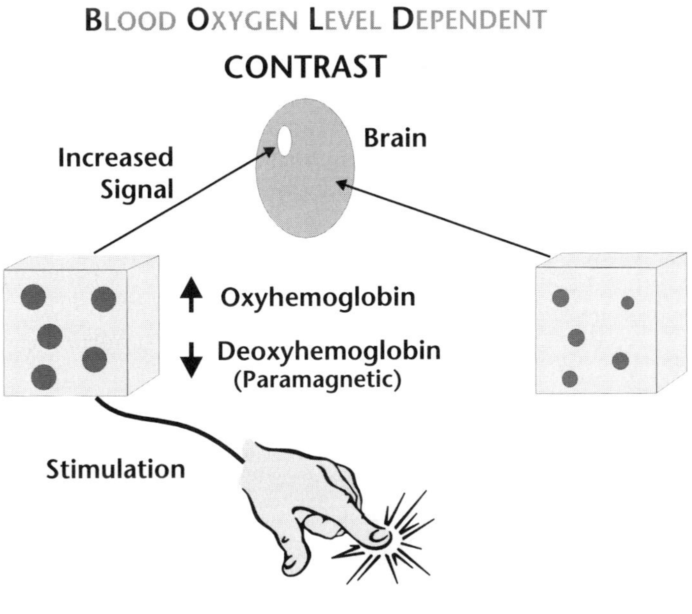

Figure 13-4. The BOLD effect for visualizing areas of the brain activated by stimulation.

that it is not invasive and can be used continuously to observe dynamic effects.

BOLD is based on the fact that deoxyhemoglobin is paramagnetic but oxyhemoglobin is not. This means that deoxyhemoglobin has a higher magnetic susceptibility and produces more local field inhomogeneities than oxyhemoglobin does. This results in more rapid proton dephasing in areas of high deoxyhemoglobin concentration.

The BOLD technique is used to visualize activated areas in the brain through a series of events as illustrated in Figure 13-4. Brain activity produces an increase in metabolism and oxygen consumption. The associated vasodilatation results in an increase in blood flow. When the flow (oxygen delivery) increases more than the oxygen consumption, there is an increase in the local oxygen level. The increase in oxyhemoglobin in relationship to the deoxyhemoglobin reduces the susceptibility effect and rate of proton dephasing. This results in an increase in the local T2* value.

When imaged with a T2*-weighted acquisition, activated brain areas will produce an increased signal intensity and brightness. The increase in signal intensity is small, on the order of a few percent. However, by taking the difference between images acquired *with* and *without* activation, the areas of activation can be observed.

Perfusion Imaging

Blood flow through the microvasculature, or tissue perfusion, can be evaluated with MRI.

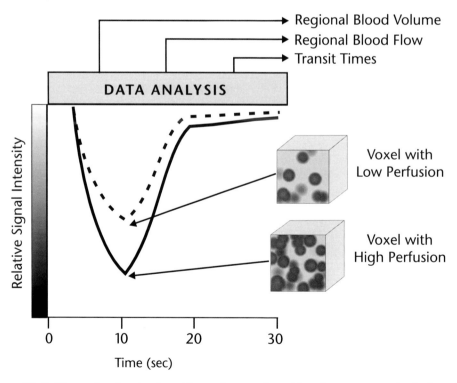

Figure 13-5. The reduction in signal intensity produced by the passage of a bolus of contrast media provides information on tissue perfusion.

MAGNETIC RESONANCE IMAGING

In the previous chapter we learned that when blood is flowing through relatively large vessels, compared to capillaries, there are several physical effects (time effects and phase changes) that can be used as sources of contrast and vessel visualization. These effects do not work for the small capillaries that distribute the blood throughout the tissues. Therefore, other methods for producing contrast must be used. The objective is not to visualize the flow in individual vessels, as in angiography, but to evaluate the combined flow through many microvessels within a volume of tissue such as a voxel.

A common method for imaging perfusion is illustrated in Figure 13-5. The process is started by administering a bolus of contrast media to the patient. As the bolus passes through a specific tissue, there will be an increase in local magnetic susceptibility. This produces a decreased signal intensity in T2*-weighted images. Voxels with the highest perfusion will experience the largest reduction in signal intensity as the bolus passes through.

The reduction in signal intensity with the passage of the bolus through two different tissues is illustrated in Figure 13-5. Acquisition of perfusion data requires fast imaging methods, such as EPI, because images must be acquired every few seconds to properly measure the characteristics of the bolus passage.

The area under the signal versus time curve is proportional to the regional blood volume. By measuring the transit time of the bolus information on the regional blood, flow can be obtained.

Mind Map Summary
Functional Imaging

INTRAVOXEL DEPHASING

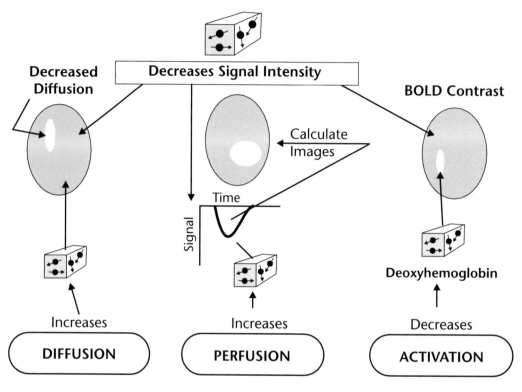

Diffusion imaging is based on the principle that the diffusion motion of molecules produces intravoxel dephasing when appropriate gradients are applied. The diffusion sensitizing gradients can be adjusted to produce different levels of diffusion sensitivity, making it possible to measure the apparent diffusion coefficient values in voxels of tissue. These values can be displayed as a map image. A diffusion-weighted image can also be produced in which areas with decreased diffusion will appear bright.

Perfusion imaging is achieved by injecting a bolus of contrast media. As the bolus passes through the imaged tissue, the intravoxel dephasing is increased and the signal intensity drops. When the data is collected as a rapid sequence of dynamic images, several different perfusion parameters can be calculated and displayed as images.

BOLD is an imaging method that can be used to visualize activated areas within the brain. The contrast is produced by a shift in the deoxyhemoglobin-oxyhemoglobin ratio resulting from the vasodilatation in the activated area. The signal intensity in the activated area, with increased oxyhemoglobin, increases because deoxyhemoglobin is a paramagnetic substance that contributes to intravoxel dephasing.

14

Image Artifacts

Introduction And Overview

There are a variety of artifacts that can appear in MR images. There are many different causes for artifacts, including equipment malfunctions and environmental factors. However, most artifacts occur under normal imaging conditions and are caused by the sensitivity of the imaging process to certain tissue characteristics such as motion and variations in composition.

There are many techniques that can be applied during the image acquisition process to suppress artifacts. In this chapter we will consider the most significant artifacts that degrade MR images and how the various artifact suppression techniques can be employed.

An artifact is something that appears in an image and is not a true representation of an object or structure within the body. Most MRI artifacts are caused by errors in the spatial encoding of RF signals from the tissue voxels. This causes the signal from a specific voxel to be displayed in the wrong pixel location. This can occur in both the phase-encoding and frequency-encoding directions, as shown in Figure 14-1. Errors in the phase-encoding direction are more common and larger, resulting in bright streaks or ghost images of some anatomical structures. Motion is the most common cause, but the aliasing effect can produce ghost images that fold over or wrap around into the image.

ARTIFACTS

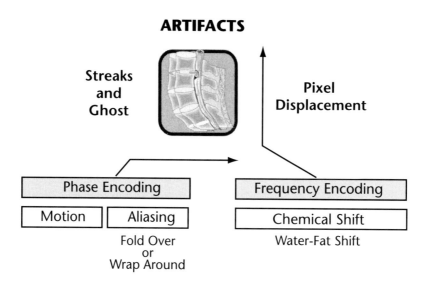

Streaks and Ghost

Pixel Displacement

Phase Encoding

Motion	Aliasing

Fold Over
or
Wrap Around

Frequency Encoding

Chemical Shift

Water-Fat Shift

Figure 14-1. Classification of the most common MRI artifacts.

Errors in the frequency-encoding direction are limited to a displacement of just a few pixels that can occur at boundaries between fat and nonfat tissues in which most of the protons are contained in water.

Motion-Induced Artifacts

Movement of body tissues and the flow of fluid during the image acquisition process is the most significant source of artifacts. The selection of a technique that can be used to suppress motion artifacts depends on the temporal characteristic of the motion (periodic or random) and the spatial relationship of the moving tissue to the image area. Figure 14-2 shows the types of motion that can produce artifacts and techniques that can be used to reduce them.

At this point we need to make a clear distinction between blurring and artifacts. We recall that the principal source of blurring in MRI is the size of the individual voxels. Under some imaging conditions motion can produce additional blurring, which reduces visibility of

detail and gives the image an unsharp appearance. This is especially true for motion that causes a voxel to change location from one acquisition cycle to another. Blurring occurs when the signals from an individual voxel come from the different locations occupied by the voxel. The signals are smeared over the region of movement. Artifacts, or ghost images, occur when the signals are displayed at locations that were never occupied by the tissue.

Most of the motion-induced artifacts are produced by dephasing or phase errors. We recall from Chapter 12 that flow and other forms of motion can produce both intravoxel and intervoxel phase problems. Intravoxel dephasing generally results in reduced signal intensity, whereas intervoxel phase errors produce artifacts in the phase-encoded direction. The phase-encoding process is especially affected by body motion, which causes a particular anatomical structure to be in a different location from one acquisition cycle to another. This contributes to the phase error and to the production of artifacts.

SOURCE OF MOTION ARTIFACTS

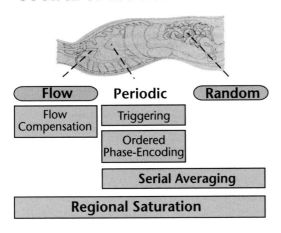

Figure 14-2. Various types of motion that produce artifacts in MR images and correction techniques.

Phase-Encoded Direction

Motion artifact streaks and ghosting always occur in the phase-encoded direction. Prior to an acquisition the operator can select which direction in the image is to be phase-encoded as opposed to frequency-encoded. This makes it possible to place the artifact streaks in specific directions. This is a very helpful technique for protecting one anatomical area from motion and flow artifacts produced in another area. It does not eliminate the artifacts but orients them in a specific direction.

Cardiac Motion

Cardiac activity is a source of motion in several anatomical locations. The movement of the heart produces artifact streaks through the thoracic region and blurring of cardiac structures when the heart is being imaged. In other parts of the body, pulsation of both blood and cerebrospinal fluid (CSF) can produce artifacts and loss of signal intensity.

Triggering

Synchronizing the image acquisition cycle with the cardiac cycle is an effective technique for reducing cardiac motion artifacts. An EKG (electrocardiogram) monitor attached to the patient provides a signal to trigger the acquisition cycle.

The R wave is generally used as the reference point. The initiation of each acquisition cycle is triggered by the R wave. Therefore, an entire image is created at one specific point in the cardiac cycle. This has two advantages. The motion artifacts are reduced and an unblurred image of the heart can be obtained. The delay time between the R wave and the acquisition cycle can be adjusted to produce images throughout the cardiac cycle. This is typically done in cardiac imaging procedures.

Maximum artifact suppression by this technique requires a constant heart rate. Arrhythmias and normal heart-rate variations reduce the effectiveness of this technique.

Cardiac triggering is also useful for reducing artifacts from CSF pulsation. This can be especially helpful in thoracic and cervical spine imaging.

Flow Compensation

The technique of flow compensation or gradient moment nulling was described in Chapter 12. In addition to compensating for blood flow effects, this technique can be used to reduce problems arising from CSF pulsation, especially in the cervical spine. It actually provides two desirable effects. The rephasing of the protons within each voxel increases signal intensity from the CSF, especially in T2-weighted images. It also reduces the motion artifacts.

Respiratory Motion

Respiratory motion can produce artifacts and blurring in both the thoracic and abdominal regions. Several techniques can be used to suppress these motion effects.

Averaging

The technique of signal averaging is used primarily to reduce signal noise, as described in Chapter 10. However, averaging has the additional benefit of reducing streak artifacts arising from motion. If a tissue voxel is moving at different velocities and in different locations during each acquisition cycle, the phase errors will be different and somewhat randomly distributed. Averaging the signals over several acquisition cycles produces some degree of cancellation of the phase errors and the artifacts. There are several different ways that signals can be averaged. Serial rather than parallel averaging gives the best artifact suppression. Serial averaging is performed by repeating two or more complete acquisitions and averaging. Parallel averaging is performed by repeating an imaging cycle two or more times for each phase-encoded gradient step. With serial averaging there is a much longer time between the measurements made at each phase-encoded step. This gives a more random distribution of phase errors and better cancellation. As with noise, increasing the number of signals averaged (NSA) reduces the intensity of artifacts but at the cost of extending the acquisition time. The averaging process reduces artifacts but not motion blurring.

Ordered Phase-Encoding

An artifact reduction technique that is used specifically to compensate for respiratory motion is ordered phase-encoding. We recall that a complete acquisition requires a large number of phase-encoded steps. The strength of the phase-encoded gradient is methodically changed from one step to another. In a normal acquisition the gradient is turned on with maximum strength during the first step and is gradually decreased to a value of zero at the midpoint of the acquisition process. During the second half the gradient strength is increased, step by step, but in the opposite direction. The basic problem is that two adjacent acquisition cycles might catch a voxel of moving tissue in two widely separated locations. The location is also somewhat random from cycle to cycle. This contributes to the severity of the artifacts.

Ordered phase-encoding is a technique in which the strength of the gradient for each phase-encoded step is related to the amount of tissue displacement at that particular instant. This requires a transducer on the patient's body to monitor respiration. The signals from the transducer are processed by the computer and used to select a specific level for the phase-encoded gradient.

REGIONAL SATURATION

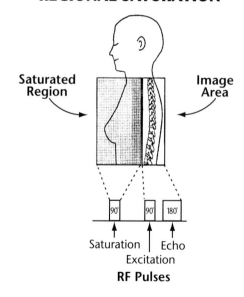

Figure 14-3. The use of regional presaturation to reduce motion artifacts.

Regional Presaturation

Regional presaturation is a technique that has several different applications. In Chapter 12 we saw how it could be used with flowing blood to eliminate the signal and produce a black-blood image. An application of this technique to reduce respiratory and cardiac motion artifacts in spine imaging is shown in Figure 14-3. With this technique a 90° RF saturation pulse is selectively applied to the region of moving tissue. This saturates or reduces any existing longitudinal magnetization to zero. This is then followed by the normal excitation pulse. However, the region that had just experienced the saturation pulse is still demagnetized (or saturated) and cannot produce a signal. This region will appear as a black void in the image. It is also incapable of sending artifact streaks into adjacent areas.

Flow

Flow is different from the types of motion described above because a specific structure does not appear to move from cycle to cycle. This reduces the blurring effect, but artifacts remain a problem.

The flow of blood or CSF in any part of the body can produce artifacts because of the phase-encoding errors. Several of the techniques that have already been described can be used to reduce flow-related artifacts.

Regional Presaturation

Regional presaturation, as described in Chapter 12, is especially effective because it turns the blood black. Black blood, which produces no signal, cannot produce artifacts.

Figure 14-4 illustrates the use of presaturation to reduce flow artifacts. The area of saturation is located so that blood flows from it into the image slice.

Flow Compensation

Flow compensation is useful when it is desirable to produce a bright-blood image. It both reduces intervoxel phase errors, the source of

REGIONAL SATURATION

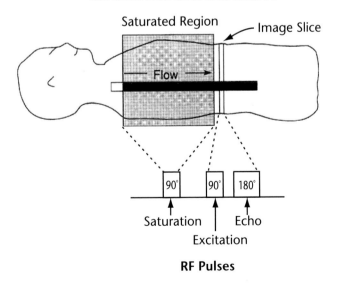

Figure 14-4. The use of regional saturation to reduce flow artifacts.

streaking, and restores some of the intravoxel magnetization and signal intensity.

Aliasing Artifacts

Aliasing, which produces foldover or wraparound artifacts, can occur when some part of the patient's body extends beyond the selected field of view (FOV). The anatomical structures that are outside of the FOV appear to "wrap around" and are displayed on the other side of the image, as shown in Figure 14-5. This occurs because the conventional imaging process does not make a sufficient number of signal measurements or samples. Because of this undersampling the anatomical structures outside of the FOV produce signals with the same frequency and phase characteristics of structures within the image area. This phenomenon is known as aliasing because structures outside of the FOV take on an alias in the form of the wrong spatial-encoding characteristics.

Two techniques that can be used to eliminate wraparound artifacts are illustrated in Figure 14-5. One procedure is to increase the size of the acquisition FOV and then display only the specific area of interest. The FOV is extended by increasing the number of voxels in that direction. This is described as oversampling. Under some conditions these additional samples or measurements will permit a reduction in the NSA so that acquisition time and signal-to-noise is not adversely affected by this technique.

An alternative method of eliminating wraparound or foldover artifacts is to apply presaturation pulses to the areas adjacent to the FOV. This eliminates signals and the resulting artifacts.

FOLDOVER ARTIFACTS

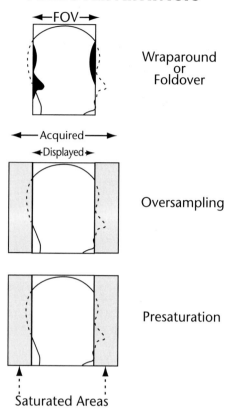

Figure 14-5. The wraparound or foldover artifact and methods for suppressing it.

Chemical-Shift Artifacts

The so-called chemical-shift artifact causes a misregistration or pixel displacement between water and fat tissue components in the frequency-encoded direction, as shown in Figure 14-6. The problem occurs because the protons in water and fat molecules do not resonate at precisely the same frequency. The shifting of the water tissue components relative to fat can produce both a void and regions of enhancement along tissue boundaries.

There are several factors that determine the number of pixels of chemical shift. Knowledge of these factors can be used to predict and control the amount of chemical shift that will appear in a clinical image.

Field Strength

We recall from Chapter 3 that the chemical shift or difference in resonant frequency between water and fat is approximately 3.3 ppm. This is the amount of chemical shift expressed as a fraction of the basic resonant frequency. The product of this and the proton resonant frequency of 64 MHz (at a field strength of 1.5 T) produces a chemical shift of 210 Hz. At a field strength of 0.5 T the chemical shift will be only 70 Hz. The practical point is that chemical shift increases with field strength and is generally more of a problem at the higher field strengths.

Bandwidth

In the frequency-encoded direction the tissue voxels emit different frequencies so that they can be separated in the reconstruction process. The RF receiver is tuned to receive this range of frequencies. This is the bandwidth of the receiver. The bandwidth is often one of the adjustable protocol factors. It can be used to control the amount of chemical shift (number of pixels), but it also has an effect on other characteristics such as signal-to-noise.

In Figure 14-6 we assume a bandwidth of 16 kHz. If the image matrix is 256 pixels in the frequency-encoded direction, this gives 15 pixels per kHz of frequency (256 pixels/16 kHz). If we now multiply this by the chemical shift of 0.210 kHz (210 Hz), we see that the chemical shift will be 3.4 pixels.

CHEMICAL-SHIFT ARTIFACT

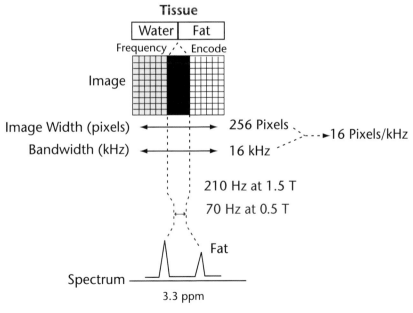

Figure 14-6. The chemical-shift artifact is related to receiver bandwidth.

The amount of chemical shift in terms of pixels can be reduced by increasing the bandwidth. This works because the chemical shift, 210 Hz, is now a smaller fraction of the image width and number of pixels.

On most MRI systems the water-fat chemical shift (number of pixels) is one of the protocol factors that can be adjusted by the operator. On some systems it is designated as bandwidth. When a different value is selected the bandwidth is automatically changed to produce the desired shift. Even though the chemical-shift artifact can be reduced by using a large bandwidth, this is not always desirable. When the bandwidth is increased, more RF noise energy will be picked up from the patient's body and the signal-to-noise relationship will be decreased. Therefore, a bandwidth or chemical-shift value should be selected that provides a proper balance between the amount of artifact and adequate signal-to-noise.

Fat suppression is useful for reducing the chemical shift artifact.

Mind Map Summary
Image Artifacts

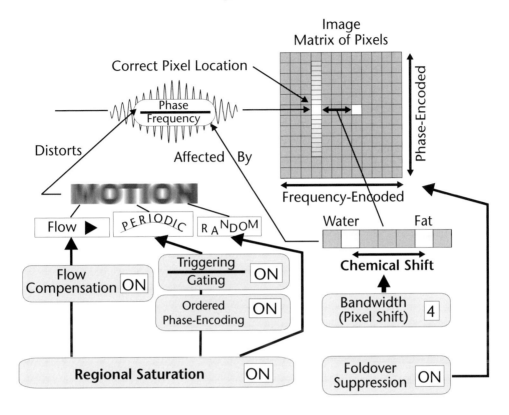

There are a variety of artifacts that can appear in MR images. Most artifacts occur when the signal from a specific tissue voxel is not displayed in the correct image pixel location. It is possible for the signal to be displaced in both the phase-encoded and frequency-encoded directions.

Motion of the tissue or fluid produces streaks or ghost images of a pixel in the phase-encoded direction. This occurs because when motion is present, a specific tissue voxel is in different locations during the phase-encoding process from one imaging cycle to another. This produces errors in the phase encoding. We can think of signals as being given an incorrect phase address. There are several techniques that can be turned on during an image acquisition to reduce the motion artifacts.

The chemical shift artifact causes signals from fat to be displaced or shifted with respect to the tissue water signals in the frequency-encoded direction. This happens because the protons in water and fat resonate at slightly different frequencies. The amount of shift (number of pixels) in an image can be controlled by adjusting the RF receiver bandwidth. However, increasing the bandwidth to reduce the chemical shift artifact results in an increase in image noise.

Foldover is a type of artifact that can occur when the anatomical region is larger than the imaged region. The image of the tissue outside the imaged area can be folded over and appear as an artifact in the image. There are techniques that can be turned on to suppress this type of artifact.

15

MRI Safety

Introduction And Overview

MRI is not an inherently dangerous process for either the patient or the MRI staff. However, the procedure does produce a physical environment and uses forms of energy that can produce injury or discomfort if not properly controlled. The potential sources of injury are not direct biological effects but interactions of the magnetic field with other objects, which in turn, might produce injury, and the application of RF energy to the patient's body, which produces heat. During an imaging procedure a patient is subjected to:

- A strong magnetic field
- Magnetic field gradients (that are rapidly changing with time)

- RF energy
- Acoustic noise (produced by the gradients).

There are potential hazards associated with each of these if certain levels are exceeded or certain conditions exist.

The purpose of this chapter is to provide the MRI staff with an understanding of the conditions that can produce injury or discomfort and to identify the actions to take (safety procedures) to produce a safe and comfortable examination.

Magnetic Fields

As we have learned, a patient must be placed in a strong magnetic field in order to perform the imaging procedure. The magnetic field is capable of producing several effects that can lead to unsafe conditions. Before considering the potential hazards, let us review the basic characteristics of a magnetic field.

Physical Characteristics

The general characteristics of a magnetic field were described in Chapter 2. Here we will review those characteristics that relate to safety.

The strength of a magnetic field is a characteristic that must be considered. Most clinical imaging is performed with magnetic fields in the general range of 0.15 T to 1.5 T, but there are some magnets with higher field strengths now being used for research and for some clinical imaging applications. As we will see later, certain safety-related effects are dependent on field strength.

Let us recall that there are two general areas of a magnetic field. One is the strong and relatively uniform field area within the bore of the magnet where the patient is located. The other is the external and somewhat weaker field that surrounds the magnet. This external field varies in strength with location. It is strongest where it meets the internal field and gradually becomes weaker with increased distance from the magnet. This variation in strength is a natural gradient in the field. There is no precise point at which the field ends. For all safety related purposes, the location at which the field strength drops to 5 gauss (the 5 gauss line) is considered to be the outer boundary of the field. Undesirable effects can be produced in both the internal and external field areas.

Magnetic fields can produce both mechanical and electrical effects on objects or materials located in the field. Both are potential sources of unsafe conditions.

Biological Effects

A major question in MRI today concerns the potential of undesirable effects of a magnetic field on the human body. At this time, there are no indications of any irreversible *biological effects* produced by the magnetic fields used for general clinical imaging. However, in many minds there is not a feeling of complete safety. A magnetic field, like x-radiation, is an invisible environment that carries with it a certain mystique in relation to biological effects. In the early days of x-ray imaging (over a century ago), a false sense of security led to unsuspected injury to persons receiving high levels of exposure. It has taken many years for us to develop a reasonably good understanding of the effects of x-ray and other forms of ionizing radiation. It is this experience that causes many to ask: How safe is exposure to magnetic fields? Are there unknown biological effects from a magnetic field that are going to appear sometime in the future?

First, we must recognize that there is a major physical difference between magnetic fields and ionizing radiation. A static magnetic field is not a form of energy that is directly transferred to human tissue like the various forms of radiation. We can think of it more as an environment somewhat like gravity. There is no basis for assuming that the undesirable effects produced by exposure to high levels of ionizing radiations (such as burns, mutations, and cancer-induction) will also apply to magnetic fields.

Magnetic fields, like gravity, do produce some effects, although they are distinctly different from the effects produced by radiation. The observed short-term effects of magnetic fields have occurred at relatively high field strengths beyond the general range of most clinical systems.

Biological effects arising from physical agents generally are of two forms: deterministic and stochastic. Deterministic effects are those that generally occur and are often observable at or shortly after the exposure; it is the severity of the effect that increases with the strength of the physical agent. Burns would be an example. A stochastic effect is one that is not necessarily produced by every exposure but where there is a statistical probability of the effect occurring. Generally, it is the probability or the risk of the effect occurring that increases with the level of exposure. The induction of mutations and cancer are the two primary stochastic effects associated with exposure to ionizing radiation. These effects have not been demonstrated to occur with exposure to magnetic fields.

Internal Objects

If a patient's body contains internal ferromagnetic objects, there is a potential for injury when the patient is placed within the magnetic field. As we know, a magnetic field exerts a force or torque on ferromagnetic objects. The pulling and twisting on objects within the body can tear tissue, rupture blood vessels, and produce fatal injuries.

The most common types of objects that must be considered are the medically implanted ones such as prosthetic devices and surgical clips. In addition, there is the possibility of a person having embedded bullets, shrapnel from war wounds, swallowed objects, or pieces of metal embedded during an accident. Even very small metal fragments or particles can produce serious injury if they are located in a sensitive site such as the eye. There is a well-known case of a metal worker who lost his sight because of a small metal fragment that was under his eyelid and ruptured his eyeball when he was placed in a magnetic field for an imaging procedure.

External Objects

Ferromagnetic objects that are brought into the external magnetic field will experience a force that either attracts them to the sides of the magnet or propels them into the bore.

> ### Safety Procedures
>
> **The principal safety procedure to safeguard against injury from internal objects is to screen patients by interview and review of medical records to determine if there are internal objects from prior surgeries or injuries.**
>
> **The next step is to determine if any detected objects are potential hazards. This is sometimes difficult. The only assurance is when a specific type of device or object (such as a surgical clip) has been evaluated and found to be safe. Current information on many such devices is generally available on the World Wide Web.**

Objects are pulled along the natural field gradients, from the weak to the strong areas of the field. This produces the so-called projectile effect where the objects are accelerated to high velocities and can injure anyone in their path, typically the patient.

This can occur with objects of all sizes. Accidents have occurred with large objects such as forklift tongs, metal chairs, and mop buckets. Small objects such as medical instruments carried in pockets and hairpins can easily become flying projectiles.

Magneto-electrical Effects

One of the laws of physics is that an electrical current will be induced or generated in a conducting material that is moved through a magnetic field, or is located in a changing magnetic field. In fact, this is the principle of the electric generator and the transformer. There is a possibility that motion within the body, such as cardiac activity, will generate some internal electrical currents. However,

Safety Procedures

The principal safety procedure is proper training of all persons who enter the room containing the magnet.

Signs should be posted informing patients and staff about the presence of a strong magnetic field and the associated hazards.

The imaging staff should monitor all other persons who enter the room and who might bring hazardous objects into the room.

The magnet room should be properly secured when not under the direct observation and supervision of the imaging staff.

Special attention should be given to ensure the safety of cleaning and maintenance staff who might not have adequate safety training and who often work during non-operational hours with minimum supervision.

these currents do not appear to produce any undesirable effects, such as fibrillation, at the field strengths used for clinical imaging.

The most common effect of this type is an elevated T wave in the EKG signal from a patient being monitored in the magnetic field. This is the time within the cardiac cycle when the blood flow velocity is the highest. A voltage is generated by this flow through the magnetic field that adds to the normal T wave, making it appear to be enhanced. It does not represent a change in cardiac activity.

Activation of or Damage to Implanted Devices

Some implanted electronic devices, such as cardiac pacemakers, have magnetically activated switches. The reason for this is so physicians can temporarily turn off the pacemaker for testing by placing a small magnet on the surface of the patient near the pacemaker site. The concern is that if a patient with this type of pacemaker should enter a magnetic field, the pacemaker would be turned off, which could be potentially fatal. This type of activation could also occur in the external field surrounding a magnet. This is why it is a common

practice to exclude patients with pacemakers from the field areas within the 5 gauss line unless it has been determined that the field will not have an adverse effect on their device.

There is a possibility that some implanted electronic devices could be damaged by the high field strength within the magnet.

Safety Procedures

Exclude persons with implanted electronic devices from the magnetic field unless it has been determined that there is no adverse effect on a particular device.

Gradients

We recall that a gradient is a non-uniformity in the magnetic field produced by turning on the various gradient coils. A gradient is a variation in field strength over space, which in itself is not a safety problem. However, potential problems are produced at the times when the gradients are being turned on and off. During these times the magnetic field is changing rapidly with respect to time. A time-varying magnetic field induces or generates electrical

Safety Procedures

It is assumed that gradient design is within appropriate safety limits.

currents through conductive materials located within the changing field, as illustrated in Figure 15-1. This includes human tissue as well as any implanted materials or devices.

For a specific material or device the amount of current produced depends on the rate at which the magnetic field is changed. This is expressed as the factor, dB/dt, where B is the symbol for field strength. This is in the units of tesla per second. The U.S. Food and Drug Administration (FDA) provides guidelines on dB/dt values that can be considered to be safe operating levels. A maximum value of 6 T/sec applies to an imaging system, but other, higher values apply to specific imaging conditions.

All MR systems do not have gradients that change at the same rate (dB/dt). This is actually the gradient slew rate described in Chapter 2. The trend in equipment development is to go to higher values to perform certain types of acquisitions, such as echo planar imaging.

RF Energy

During an imaging procedure pulses of RF energy are transmitted to the patient's body. Most of the energy is absorbed by the tissue and is converted to heat that has the potential of increasing the temperature within the body. This is the physical principle of the microwave oven. In the microwave oven high-frequency radio energy (microwaves) are transmitted at a high power level to a relatively small mass of food. This high concentration of energy can raise the temperature by hundreds of degrees.

MAGNETIC FIELD GRADIENTS

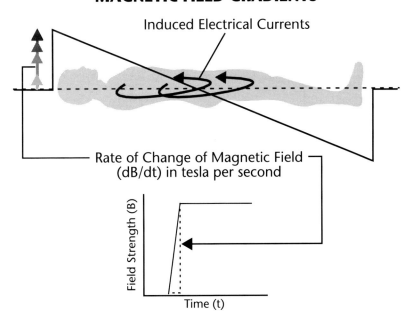

Figure 15-1. Electrical currents are generated by rapidly changing magnetic fields (the gradients).

In MRI the RF energy applied to a patient's body is at a much lower power and is distributed over a larger mass of tissue than in the microwave oven example. The elevation in temperature produced during MRI depends on many factors. Variations among patients that will affect this include body size and blood circulation, which distributes heat and promotes cooling.

The principal issue associated with an imaging procedure is the rate at which the energy is transferred to the body. The factors that determine this are illustrated in Figure 15-2. Recall that the rate of energy transfer is the physical quantity *power* and is expressed in the units of *watts*. The rate at which temperature is increased is related to the concentration of power in tissue.

Specific Absorption Rate (SAR)

In MRI, the rate at which energy (heat) is deposited in a unit mass of tissue is expressed in terms of the *Specific Absorption Rate* (SAR) in units of watts per kilogram of tissue. For specific tissue conditions, the rate of temperature rise will be proportional to the SAR. The total increase in temperature is determined by the SAR, the duration of the exposure to the RF pulses, and various tissue cooling and heat distribution factors.

Limits have been established that are thought to represent insignificant risk to patients if the examination times are within established ranges. These are shown in Figure 15-2.

The SAR is determined by a combination of factors associated with the MR system and

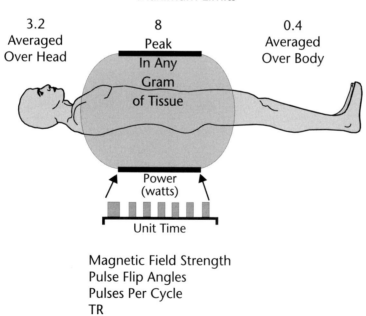

Figure 15-2. The factors that determine the SAR, or rate of heat production during an image acquisition.

the acquisition protocol as listed in Figure 15-2. Each RF pulse delivers a quantity of energy to the body. Therefore, the SAR is determined by the energy in each pulse and the rate at which the pulses are applied.

Magnetic Field Strength

The field strength is a factor because the amount of RF energy required to produce a pulse with a specific flip angle increases with field strength. Therefore, a 90° pulse delivers more energy in a 1.5 T field than in a 0.5 T field.

Pulse Flip Angles

The energy of a pulse increases with flip angle. A 180° pulse delivers more energy than a 90° pulse.

Pulses per Cycle

We have observed that there are many combinations of RF pulses used by the different imaging methods, or pulse sequences. A simple spin echo acquisition might use just two pulses (90° and 180°), while other methods, such as fast or turbo spin echo acquisitions use many 180° pulses. This produces more energy delivered during each imaging cycle.

TR

Since TR is the duration of the imaging cycle, it affects the rate at which energy is delivered. Reducing TR concentrates the pulses into a shorter time and results in an increase in the SAR.

Determining the SAR for a Patient

We have just seen that there are many variable factors that determine the SAR for a specific patient undergoing an imaging procedure. These include both the size of the patient and the specific imaging protocol that is being used.

It would be difficult to take all of the factors and manually calculate the SAR value. Fortunately, most MR systems perform this calculation during the preparation of the patient and of the protocol. The SAR value might be displayed for the operator to visually check and to automatically prevent the system from proceeding with an imaging protocol that will exceed established SAR limits.

Surface Burns

It is possible for metal objects, such as monitoring leads and electrodes that are in contact with the patient's body to act as an antenna and to pick up some of the RF energy. There have been cases where the concentration of energy at the point of contact has produced burns.

Safety Procedures

Be aware of the potential problem and use only devices and lead configurations that have been demonstrated to be safe. Investigate any reports from the patient of discomfort.

Acoustic Noise

The noise produced by the gradients is loud, especially for high performance gradients performing certain types of acquisitions. This is often uncomfortable for patients, and can be a potential source of longer term effects to hearing.

Safety Procedures

Provide hearing protection and audio diversions as appropriate.

Mind Map Summary
MRI Safety

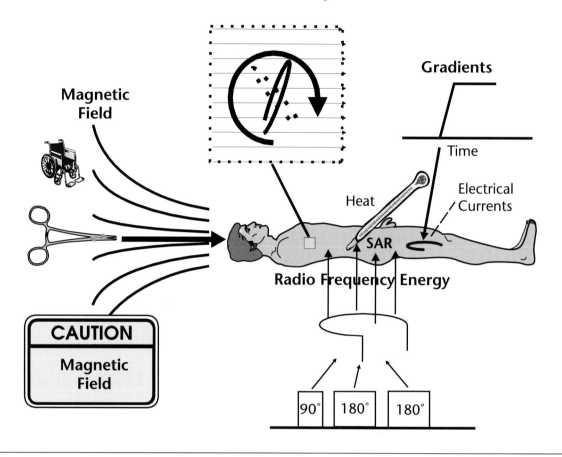

The MR imaging procedure is generally safe for both patients and staff if certain precautions are observed. However, there are three, specific physical environments that, if not properly controlled, could produce injury. They are: a strong magnetic field, rapid-changing magnetic field gradients, and RF energy.

The strong static (non-changing) magnetic field does not appear to produce any undesirable direct biological effects. However, injury can result from the effect of the magnetic field on metal objects, both external to and within the patient's body. Objects, such as surgical clips and other implanted devices, can be pulled or rotated by the magnetic field. Magnetic-susceptible objects brought into the external field can be pulled into the magnetic field and become projectiles that can injure patients.

A gradient, that is a rapidly changing magnetic field when it is being turned on and off, can induce electrical currents within a patient's body or equipment attached to the patient. This is why limits have been established on the rate of change for the gradient fields.

The RF energy applied to a patient's body is absorbed and converted to heat. This can increase tissue temperatures. The SAR is the quantity that expresses the rate at which energy is being deposited. SAR is determined by several factors: the number of pulses in a time interval, the flip angles of the pulses, and the strength of the magnetic field. Limits have been established to minimize adverse effects.

Index

C

Calcium, 27
Cancer, 157
Capillaries, 143
Carbon, 27, 28
Cardiac activity, 157
Cardiac motion, 147
    flow compensation, 147
    triggering, 147
Cerebrospinal fluid (CSF), 39, 40, 147
Chelation, 46
Chemical elements, 27
Chemical shift, 3-4, 32, 100
Chemical-shift artifacts, 100, 150-52
Circularly polarized mode, 21
Claustrophobia, 16
Coils
    gradient, 17
    quadrature, 21
    radio frequencey (RF), 20, 28
Components. *See* System components
Computed tomography (CT) imaging, 8
Computer functions, 21-22
Computers, 19, 106
    image reconstruction, 99-100
Contrast
    low, 75
    mixed, 76
Contrast agents, 45
    diamagnetic material, 44, 45
    ferromagnetic material, 19-20, 44, 46
    negative, 46
    paramagnetic material, 44, 45-46
Contrast angiography. *See* Contrast enhanced
    angiography; In-flow contrast angiography;
    Phase contrast angiography
Contrast enhanced angiography, 131, 135
Contrast enhancement, 75-76
Contrast sensitivity, 7-8, 9, 53-57, 73-75, 119
Contrast (windowing), 22
Contrast-detail curve, 104
Control of image quality characteristics, 7-11
Coolants, 16
Copper, 15, 16
Coronal slice orientation, 92
Crystalline structure, 46
CSF. *See* Cerebrospinal fluid
CT. *See* Computed tomography (CT) imaging
"Curtain of invisibility," 103, 104
Cycle repetition time. *See* Time of repetition (TR)

D

D. *See* Diffusion coefficient
dB/dt values, 159
Deoxyhemoglobin, 142
Detail (blurring), 8-9, 10, 11, 103-12, 146
    See also Image detail and noise
Deterministic biological effects, 156-57
Diamagnetic material, 44, 45
Diffusion, 3
Diffusion coefficient (D), 138
Diffusion direction, 139
Diffusion imaging, 137-41
Diffusion sensitivity, 140-41
Diffusion-weighted images, 141
Disease, 1
Display of images, 22, 25
Doppler ultrasound image, 130

E

Earth's magnetic field, 15
Echo event and signals, 53
Echo planar imaging (EPI) method, 76-77, 120,
    143, 159
Echo planar imaging (EPI) speed factor, 77
Echoes, 28
Eddy currents, 18
EKG. *See* Electrocardiogram
Electric generators, 157
Electric motors, 21
Electrical conductors, 106
Electrical power, 16
Electrically conducted enclosure, 21
Electrocardiogram (EKG), 147, 158
Electronic equipments, 19
Electrons, unpaired, 46
Electrosurgery units, 106
Elevated T wave, 158
Embedded bullets, 157
EPI method. *See* Echo planar imaging method
Equipment, 19, 21
Equipment operator, 3, 7, 161
Even-echo rephasing in vascular imaging, 129-30
Excitation, 29-30, 31, 36, 38, 53
    selective, 91, 92
Excitation/saturation-pulse flip angle, 71-73, 74
Excited state, 29, 36
External field, 18, 19
External objects and safety, 157

F

Fast acquisition methods, 120-21, 143
Fast spin echo, 120
Fat tissue, 32, 39, 40, 100, 146
   magnetization transfer in, 86
   suppression of images, 81, 82-83
   and water chemical shift, 152
FDA. *See* United States Food and Drug
  Administration
Ferromagnetic material, 19-20, 44, 46
Fibrillation, 158
Field direction, 14-15
Field strength, 14, 15, 31, 106-7
   in chemical-shift artifacts, 151
Field uniformity. *See* Homogeneity
Field of view (FOV), 105, 106, 109, 116, 150
Flip angles, 36, 61
   excitation/saturation-pulse, 71-73, 74
Flow compensation, 129
   in cardiac motion, 147
   vascular imaging, 129
Flow dephasing in vascular imaging, 129
Flow direction in phase contrast angiography,
  132-33
Flow in motion-induced artifacts, 149-50
Flow-related enhancement (bright blood) in
  vascular imaging, 124-25
Flow-void effect in vascular imaging, 125-26
Fluid movement and image types, 3
Fluids, 39
   magnetization transfer in, 86
   suppression of images, 81, 83-84
   See also Water
Fluorescent lights, 21, 106
Fluorine, 27, 28, 31
Foldover artifacts, 105
Fourier transformation, 22, 91, 98, 100, 130
FOV. *See* Field of view
Free induction decay (FID), 60, 70, 71, 76
Free proton pool, 85, 86
Free states, 40
Frequency, 91
Frequency-encoding, 89, 94-96
Frequency-encoding direction in artifacts, 145,
  146
Frequency-encoding gradient, 98
Functional imaging, 18, 137-44
   blood oxygenation level dependent (BOLD)
    contrast, 141-42
   diffusion imaging, 137-41
    ADC map images, 141
    apparent diffusion coefficient (ADC), 138-39

    Brownian motion and diffusion, 137-38
    diffusion coefficient (D), 138
    diffusion direction, 139
    diffusion sensitivity, 140-41
    diffusion-weighted images, 141
    imaging process, 139-40
   perfusion imaging, 142-43

G

G. *See* Gauss
Gadolinium, 45, 46
Gadolinium contrast media, 84, 143
Gadolinium diethylene triamine penta-acetic acid
  (GaDTPA), 46
GaDTPA. *See* Gadolinium diethylene triamine
  penta-acetic acid
Gamma-ray, 20
Gauss (G), 15
Ghost images, 100, 145, 146, 147
Gradient coils, 17
Gradient cyle, 98-99
Gradient echo, 17-18, 28, 51, 53, 65, 115
Gradient echo imaging methods, 69-80
   advantages and limitations, 69
   echo planar imaging (EPI) method, 76-77
   gradient and spin echo (GRASE) method, 77-78
   magnetization preparation, 78-79
   process, 69-71
   small angle gradient echo (SAGE) method, 71-
    76, 78
    contrast enhancement, 75-76
    contrast sensitivity, 73-75
    excitation/saturation-pulse flip angle, 71-73,
     74
Gradient moment nulling, 129
Gradient pulse, 59
Gradient and spin echo (GRASE) method, 77-78,
  120
Gradient strength, 98
Gradients, 16-18, 91
   and safety, 158-59
GRASE method. *See* Gradient and spin echo
  (GRASE) method
Gray matter, 40
Gyromagnetic ratio, 31

H

Half acquisition, 116-17
Half Fourier, 116
Half scan, 116

Head coils, 20, 108
Hearing effects, 161
Hertz (Hz), 94, 151
Homogeneity, 15, 18, 85
 See also Inhomogeneities
Hydrogen, 5, 27, 28, 31, 107
Hz. *See* Hertz

I

Image acquisition time, 11
Image artifacts. *See* Artifacts
Image brightness. *See* Brightness
Image characteristics, 91
Image detail and noise, 103-12, 119-20
 averaging, 109-10
 detail sources, 104-6
 field strength, 106-7
 noise sources, 106
 radio frequency (RF) coils, 108-9
 receiver bandwidth, 109, 152
 signal-to-noise considerations, 106, 107
 tissue characteristics, 107
 TR and TE, 107-8
 voxel size, 106
 See also Detail (blurring); Noise
Image format in angiography, 134-35
 maximum intensity projection (MIP), 134-35
 surface rendering, 135
Image quality, 7-11, 103, 113
Image reconstruction, 22, 49, 50, 91, 99-100
Image reconstruction phase, 89, 90, 116
Image storage and retrieval, 22
Imaging cycle, 51-53
Imaging equipment, 19
Imaging methods, 50-51
Imaging process, 49-58
 acquisition, 49-50
 contrast sensitivity, 53-57
  proton density (PD) contrast, 55-56, 57
  T1 contrast, 53-55, 57
  T2 contrast, 56-57
 imaging cycle, 51-53
  echo event and signals, 53
  excitation, 53
  time of repetition (TR), 49, 52, 53
  time to echo (TE), 52, 53, 57
 imaging methods, 50-51
 k space, 49
 protocol, 50
 reconstruction, 49, 50
Implanted objects and devices, 157, 158

In-flow contrast angiography, 133-34
 three-dimensional (3-D) volume acquisition, 133-34
 two-dimensional (2-D) volume acquisition, 134
In-flow effect, 124
In vivo spectroscopic analysis, 26
Inhomogeneities, 18-19, 60, 70, 75
 See also Homogeneity
Internal field, 18, 19
Internal objects and safety, 167
Intervoxel phase, 126, 127, 130-31, 146
 artifacts, 130-31
 phase imaging, 130
 vascular imaging, 130-31
Intravoxel phase, vascular imaging, 126, 127, 128
Inversion pulse, 79
Inversion recovery, 64-65, 79
 short time inversion recovery (STIR) fat
suppression, 82-83
Ionizing radiation, 156
Iron, 19-20, 45, 46
Iron oxide particles, 46
Isotopic abundance, 27, 28

K

k space, 49, 76, 77, 78, 98, 100, 114, 116, 121

L

Larmor frequency, 31
Linear polarized mode, 21
Liquid helium, 16
Liquid state, 27
Liver, 40
Longitudinal magnetization and relaxation, 36, 37-40, 51-52, 78-79
 See also T1 (longitudinal relaxation time)
Longitudinal relaxation time. *See* T1 (longitudinal relaxation time)
Low contrast, 75

M

Magnesium, 27
Magnetic direction, 36, 37
Magnetic field, 14-15
Magnetic field shielding, 15, 19-20
Magnetic fields and safety, 155-58
Magnetic flipping, 36-37, 38
Magnetic moment, 4, 26, 29
Magnetic muclei (protons), 4, 25-28

Magnetic relaxation times, 3, 4
Magnetic resonance imaging, 1-12, 14
    control of image characteristics, 7-11
    definition, 1
    fluid movement and image types, 3
    hydrogen, 5
    protons (magnetic nuclei), 4
    radio frequency signal intensity, 4
    spatial characteristics, 6-7
    spectroscopic and chemical shift, 3-4
    tissue characteristics and image types, 3, 5-6
    tissue magnetization, 2, 4
    See also Safety; System components
Magnetic susceptibility, 44-46
Magnetization, 4
Magnetization preparation, 78-79
Magnetization transfer, 85, 86
Magnetization transfer contrast (MTC), 85-86
Magnetization vector, 35
Magnetized, 25
Magneto-electrical effects and safety, 157-58
Magnets, 15-16
Magnification (zooming), 22
Maintenance staff, 158
Manganese, 45
Mass numbers, 27
Matrix, reduced in phase-encoded direction, 116
Matrix size, 105-6, 114, 115-16
Maximum intensity projection (MIP), 134-35
Mean (rms) distance traveled, 138
Medical equipment, 21
Megahertz (MHz), 31, 95, 151
Metal ions, 45-46
Metal objects, 19, 85, 161
MHz. *See* Megahertz
Microvasculature perfusion, 142-43
Microvessels, 143
Microwave ovens, 159
Milliseconds (msec), 18
Millitesla per meter (mT/m), 18
MIP. *See* Maximum intensity projection
Mixed contrast, 76
Modulators, 21
Molecular oxygen, 45
Molecular size, 39-40
Motion of patient, 113
Motion-induced artifacts, 86, 87, 146-50
Movement of blood, 123
MRI. *See* Magnetic resonance imaging
msec. *See* Milliseconds
mT/m. *See* Millitesla per meter
Multi-slice imaging, 92-93
Multiple spin echo, 61, 63-64

Muscle, 40
Musical instrument strings, 31
Mutations, 157

N

Nb-Ti. *See* Niobium-titanium
Negative contrast agents, 46
Neutron-proton composition, 26
Niobium-titanium (Nb-Ti), 16
Nitrogen, 27, 28
Nitroxide free radicals, 45
NMR. *See* Muclear magnetic resonance
Noise, 9, 10, 11, 103-12
    See also Image detail and noise; Visual noise
Non-ionizing radiation, 20
Normal tissues, 3
NSA. *See* Number of signals averaged
Nuclear alignment, 28-29
Nuclear magnetic effect, 44
Nuclear magnetic resonance (NMR), 14, 25-34
    chemical shift, 32
    field strength, 31
    Larmor frequency, 31
    magnetic nuclei, 25-28
        isotopic abundance, 28
        radio frequency (RF) intensity, 26-27
        relative sensitivity and signal strength, 28
        relative signal strength, 27
        spins, 26
        tissue concentration of elements, 27
    nuclear magnetic interactions, 28-30
        excitation, 29-30, 31
        nuclear alignment, 28-29
        precession and resonance, 29, 30-31
        relaxation, 30
Nuclear medicine procedures, 4
Number of signals averaged (NSA), 110, 116, 117, 120

O

Odd-echo images, 129
Optimization of procedure, 113-22
Optimized protocol development, 11, 119-20
Ordered phase-encoding in respiratory motion, 148
Orthogonal directions, 17
Oxygen, 27, 28
    molecular, 45
Oxygen level, 142
Oxyhemoglobin, 142

P

Pacemakers, 158
Paramagnetic material, 44, 45-46
Parts per million (ppm), 15, 151
Passive shielding, 19-20
Pathologic conditions, 3, 11, 27, 81, 83, 137
Patient discomfort, 161
Patient environments, 16
Patient motion, 113
Patient safety. *See* Safety
Patient's body, 106
PD. *See* Proton density (PD) images
Perfusion, 3
Perfusion imaging, 142-43
Permanent magnets, 16
Phase, 91
Phase contrast angiography, 131-33
    flow direction, 132-33
    velocity, 132
Phase effects in vascular imaging, 126-31
Phase imaging, intervoxel, 130
Phase-encoded direction in motion-induced
    artifacts, 147
Phase-encoding, 89, 96-98
Phase-encoding gradient, 98
Phased array, 20
Phosphorus, 27, 28, 31
Pixels, 6, 7, 27, 57, 99, 100, 104, 150
Polarization, 21
Post processing, 22
Postal analogy, 100
Potassium, 27
Power, 160
Power amplifiers, 21
Power monitoring circuit, 21
ppm. *See* Parts per million
Precession and resonance, 29, 30-31
Presaturaion technique, 126
Procedure optimization, 113-22
    See also Acquisition time
Projectile effect, 157
Prosthetic devices, 157
Protocol factors, 49, 52, 113, 118, 152
    interactions of, 118-19
    and safety, 161
Protocols, 50, 104
    in the acquisition process, 22
    optimization of, 11, 119-20
Proton density (PD), 3, 5, 6, 27, 52, 63
    contrast, 52, 53, 55-56, 57, 61, 62
    gradient echo methods, 71, 75, 78
Proton dephasing, 43-44

Protons (magnetic nuclei), 4
Pulse inversion, 79
Pulse sequences, 51
Pulses, 36, 38
    saturation, 78

Q

Quadrature coils, 21
Quality of image, 7-11, 103, 113
Quench, 16

R

R wave, 147
Radiation, ionizing, 156
Radio broadcast stations, 94
Radio communications devices, 21
Radio frequency (RF) coils, 20, 28, 108, 108-9
Radio frequency (RF) energy, 21, 106
    safety, 159-61
Radio frequency (RF) intensity, 26-27
Radio frequency (RF) pulse, 69
Radio frequency (RF) sequence, 61, 63
Radio frequency (RF) signal intensity, 4
Radio frequency (RF) silent condition, 41
Radio frequency (RF) system, 20-21
Radio receivers, 31
Radio signals, 25
Radio transmitters, 106
Radioactive nuclide imaging, 4
Receiver bandwidth, 109
Receiving system, 21, 106
Reconstruction of image. *See* Image reconstruction
Rectangular field of view (FOV), 116
Reduced matrix in phase-encoded direction, 116
Reduction in signal intensity formula, 140
Regional presaturation
    and flow in motion-induced artifacts, 149
    respiratory motion, 148, 149
Regional saturation, 86-87
Relative sensitivity and signal strength, 28
Relative signal strength, 27
Relaxation, 3, 30, 37, 38, 55, 59, 60
    See also Tissue magnetization and relaxation
Resistive magnets, 16
Resonance, 25, 29, 30-31
    tissue, 14
Resonant characteristic, 25
Resonant frequency, 31, 84, 95-96
Respiratory motion, 147-49
    averaging, 148

tissue magnetization, 36-37
   magnetic direction, 36, 37
   magnetic flipping, 36-37, 38
   transverse magnetization and relaxation, 36,
     37, 38, 40-44, 51-52, 59, 60
     proton dephasing, 43-44
     T2 contrast, 41-43
Tissue perfusion, 142-43
Tissue resonance, 14
Tomographic imaging process, 6
TR. *See* Time of repetition
Training for safety, 158
Transaxial slice orientation, 92
Transformers, 157
Transmitter, 20-21
Transverse magnetization and relaxation, 36, 37,
   38, 40-44, 51-52, 59, 60
   See also T2 (transverse relaxation time)
Transverse relaxation time. *See* T2 (transverse
   relaxation time)
Triggering in cardiac motion, 147
TS. *See* Time after saturation
Turbo factor, 120-21
Turbo spin echo, 120
Two dimensional (2-D) volume, 92, 94, 98, 100,
   119
Two-dimensional (2-D) volume acquisition, in-
   flow contrast angiography, 134

U

Ultrasound, 20
Ultrasound image, 130
United States Food and Drug Administration
   (FDA), 159
Unpaired electrons, 46

V

Vascular flow (angiography), 3, 6, 18, 69, 86
Vascular imaging, 123-36
   angiography, 131-35
     image format, 134-35
     in-flow contrast angiography, 133-34
     phase contrast angiography, 131-33
   phase effects, 126-31
     even-echo rephasing, 129-30
     flow compensation, 129
     flow dephasing, 129
     intervoxel phase, 126, 127, 130-31, 146
     intravoxel phase, 126, 127, 128
   selective saturation, 126

time effects, 124-26
   flow-related enhancement (bright blood),
     124-25
   flow-void effect, 125-26
   See also Angiography (vascular flow)
Vehicles, 19
Velocity in phase contrast angiography, 132
Vertical magnetic fields, 16
Viewing control, 22
Visual noise, 103
   See also Noise
Volume acquistion, 93-94
Voxel size, 104-5, 106, 109, 115, 119
Voxels, 6, 7, 11, 17, 27, 35, 39, 44, 59, 60, 70, 91,
   95, 96, 98, 99, 100

W

Warning signs for safety, 158
Water, 32, 39, 40, 100, 146
   and fat chemical shift, 152
   See also Fluid
Watts, 160
Weighted images, 53
   T1, 3, 41, 46
   T2, 3, 41-43
   See also T1; T2
White matter, 40
Windowing (contrast), 22
Work-station computers, 22
World Wide Web, 157

X

x, y, and z directions, 17
X-ray angiography, 123
X-ray imaging, 8, 20, 156

Z

Zooming (magnification), 22